T0233664

CISM COURSES AND LECTURES

Series Editors:

The Rectors of CISM
Sandor Kaliszky - Budapest
Mahir Sayir - Zurich
Wilhelm Schneider - Wien

The Secretary General of CISM
Giovanni Bianchi - Milan

Executive Editor
Carlo Tasso - Udine

The series presents lecture notes, monographs, edited works and proceedings in the field of Mechanics, Engineering, Computer Science and Applied Mathematics.
Purpose of the series in to make known in the international scientific and technical community results obtained in some of the activities organized by CISM, the International Centre for Mechanical Sciences.

INTERNATIONAL CENTRE FOR MECHANICAL SCIENCES

COURSES AND LECTURES - No. 354

SUMMATION THEOREMS IN STRUCTURAL STABILITY

EDITED BY

T. TARNAI
TECHNICAL UNIVERSITY OF BUDAPEST

Springer-Verlag Wien GmbH

Le spese di stampa di questo volume sono in parte coperte da
contributi del Consiglio Nazionale delle Ricerche.

This volume contains 106 illustrations

In order to make this volume available as economically and as
rapidly as possible the authors' typescripts have been
reproduced in their original forms. This method unfortunately
has its typographical limitations but it is hoped that they in no
way distract the reader.

ISBN 978-3-211-82704-8   ISBN 978-3-7091-2912-8 (eBook)
DOI 10.1007/978-3-7091-2912-8

PREFACE

Summation formulae are used in the theory of elastic stability so that approximate estimates of the critical load factors of a complex problem are obtained by combining the load factors of subproblems in different ways. That is, the subproblems are solved and an approximate value of the load factor of the original problem is given by the load factors of subproblems by addition. If the critical load factors are directly added, then a formula is called a Southwell type formula. If the reciprocals of the critical load factors are added, then the formula is called a Dunkerley type. The practical advantage of the summation formulae is that, for the subproblems, there are usually solutions available, or easy to determine, while for the original problem the solution would be difficult to obtain.

This book is a published form of the lecture notes prepared by lecturers for an advanced course entitled "Summation and Bounding Theorems and their Applications to Structural Stability Problems".

The aim of the course was to present the basic mathematical principles, the main theorems and formulae, and their applications to practical problems. The course presented the basic mathematical background of elastic buckling, based on relationships for eigenvalues of linear operators in Hilbert space, and the basic principles like monotonicity, convexity principles (H.F. Weinberger). The main relationships in buckling were given not only for continuous structures but for discrete structures also (H. R. Milner) using matrix formulation. One of the aims of the course was to survey the main theorems and formulae used in engineering, like Southwell theorem, Dunkerley theorem, Föppl-Papkovich theorem, Kollár conjecture, Melan theorem (T. Tarnai), and to show, where possible, the conditions under which the results are on the safe side. The Rankine-Merchant formula for elastic-plastic structures was also discussed (M.R. Horne). Emphasis was placed on the practical applications of the theoretical results, and a large number of selected examples were presented (L. Kollár).

Although the Southwell and the Dunkerley theorems are quite old, the course focused on some important new aspects: (1) In mathematics many related results have been discovered in the field of eigenvalue approximations for one-and more-parameter eigenvalue problems which practically remained unknown for engineering. (2) The mathematical results were translated to engineering language. (3) These mathematical results enable us to apply the summation theorems in ways different from those used so far, that is, to expand their circle of applications. (4) This course was the first to give a comprehensive survey of the old and new summation formulae, including also formulae for elastic-plastic structures.

I am grateful to all the co-authors for their contribution, and for their effort and enthusiasm to complete these Lecture Notes. I am also express my gratitude to all the participants in the course and to all the lecturers for the stimulating and encouraging discussions. On behalf of the co-authors I thank CISM and its staff for the excellent organization of the course and for their kind hospitality.

T. Tarnai

CONTENTS

Page

Preface

Some Mathematical Aspects of Buckling
by H.F. Weinberger 1

Stability of Discrete Systems
by H.R. Milner 39

The Rankine-Merchant Load and its Application
by M.R. Horne 111

The Southwell and the Dunkerley Theorems
by T. Tarnai 141

Practical Examples
by L. Kollár 187

SOME MATHEMATICAL ASPECTS OF BUCKLING

H.F. Weinberger
University of Minnesota, Minneapolis, MN, USA

ABSTRACT.

This is a set of lecture notes for a very short course on the mathematical foundations of buckling theory. Since the critical buckling load can usually be characterized as the largest (or smallest) eigenvalue of a variational problem, the approximation of this load is a special case of the approximation of all the eigenvalues of a variational problem. A rather detailed analysis of this problem is to be found in the author's book *Variational Methods for Eigenvalue Approximation*, CBMS Regional Conference Series in Applied Mathematics 15, SIAM, Philadelphia, 1974. We refer the interested reader to this book for more details and for references.

1. The bifurcation of some simple systems.

We begin with a simple example of a buckling problem. Suppose a rigid straight rod of length one is attached at its lower end to a torsion spring which tries to keep it in an upward vertical position, and a downward force P is applied at its upper end.

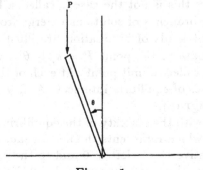

Figure 1

If θ denotes the angle of the rod with the vertical, the spring exerts a restoring torque which depends upon θ. Let us suppose for the sake of illustration that this torque is given by the function $-\theta[1 - (\theta/\pi)]$. The vertical force exerts the torque $\sin\theta$ on the lower end, and we shall neglect the gravitational force on the rod. An equilibrium position θ is any solution of the torque balance equation

$$-\theta[1 - (\theta/\pi)] + P\sin\theta = 0. \tag{1.1}$$

It is obvious that $\theta = 0$ and $\theta = \pi$ are equilibrium positions for any value of P. For $P = 0$ these are the only equilibria, and the same is still true for positive P below $\pi/4$. At this value of P there is also an equilibrium at $\theta = \pi/2$ which is a double root of the equation (1.1). As the value of P is increased further, this root splits into two roots, one of which increases with increasing P, while the other decreases. When P increases through the value 1, the greater of these equilibria increases through the value π of the upper fixed equilibrium, while the smaller equilibrium passes through the value zero of the lower fixed equilibrium. At $P = 1$ the two equilibria 0 and 1 are double roots of (1.1). For large P other roots with $|\theta| > 2\pi$ occur, but we shall not consider these here.

If we draw the equilibria for $-\pi \le \theta \le 2\pi$ as functions of P, we obtain the picture in Figure 2.

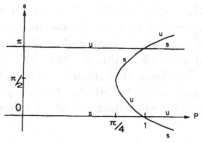

Figure 2

We note that most points on the set of equilibria have a neighborhood whose intersection with the set of equilibria is the graph of an equation of the form $\theta = \phi(P)$. An equilibrium point where this is not the case is called a **bifurcation point**, and the phenomenon of having branches of solutions emerge from such a point is called **bifurcation**. Two different kinds of bifurcation are illustrated by Figure 2. Two branches of solutions originate at the point $P = \pi/4$, $\theta = \pi/2$ and continue to the right. Such a point is often called a limit point. At each of the two bifurcation points $(1,0)$ and $(1,\pi)$ two branches of equilibria intersect. A diagram like that of Figure 2 is called a **bifurcation diagram**.

We shall be concerned with the stability of the equilibrium positions. If $P < \pi/4$, the left-hand side of (1.1), which represents the net torque, is positive for θ between the two equilibria and negative to the left of 0 and to the right of π. These torques will tend to keep the rod near the vertical position $\theta = 0$ if it starts near this position, but will make it move away from the $\theta = \pi$ equilibrium if it starts near but not at this equilibrium. We say that for $P < 1$ the equilibrium $\theta = 0$ is **stable**, while $\theta = \pi$ is **unstable**.

When $\pi/4 < P < 1$, the torque is positive for θ negative and changes its sign at each of the four equilibria. We conclude from the above reasoning that $\theta = 0$ is still stable and $\theta = \pi$ is still unstable, and that the upper of the other equilibria is stable while the lower is unstable. Similar considerations for $P > 1$ show that $\theta = 0$ is now unstable while the negative equilibrium is stable, and that $\theta = \pi$ is stable while the equilibrium above it is unstable. We have marked the stable branches in Figure 2 with the letter s and the unstable ones with the letter u.

We note that in a neighborhood of each of the bifurcation points the number of stable equilibria minus the number of unstable ones has the same value to the left and to the right of the bifurcation point. We can say, for example, that as P increases through the value 1, the equilibrium solution $\theta = 0$ loses its stability to the other equilibrium solution. This phenomenon occurs frequently, and is called an **exchange of stability**.

If we observe the rod which is initially vertical as the load P is gradually increased, we will see that, in spite of inevitable ambient noise, it tends to be remain the vertical position $\theta = 0$ until P reaches a value very slightly below 1. At this point one can expect that the noise will sooner or later drive the rod past the nearby unstable solution, so that it falls toward the other stable equilibrium, which is below the horizontal position. This phenomenon is called **buckling**. The value 1 of P at which the bifurcation occurs is called the **critical buckling load**.

We note that we can determine the stability of most equilibrium points by looking only at the derivative of the torque function on the left of (1.1) with respect to θ. An equilibrium at which the derivative is negative is stable, while an equilibrium at which the derivative is positive is unstable. We cannot draw a conclusion if the derivative is zero. The implicit function theorem implies that an equilibrium at which the derivative is not zero cannot be a bifurcation point.

We now observe that the torque on the left of (1.1) can be written as minus the derivative with respect to θ of the potential energy function

$$V(\theta; P) := \frac{1}{2}\theta^2 - \frac{1}{3\pi}\theta^3 + P\cos\theta.$$

The equilibrium points can then be characterized as those points where the derivative of V with respect to θ are zero, which are called the **critical points** of V. An equilibrium is stable if V has a local minimum there, and unstable otherwise.

These criteria can also be applied to more complicated systems. Consider, for example, a system of two rigid rods of unit length with rod number 1 connected to a fixed pivot by a spring with the same torque law as before, and the second rod connected to the other end of the first by a pivot with an identical spring which tries to keep the two rods aligned. We call the angle between the first rod and the vertical direction θ_1 and the angle between the second rod and the vertical θ_2. We suppose that a vertical force P_1 pushes downward at the top of the first rod, and that a force P_2 pushes downward at the top of the second rod.

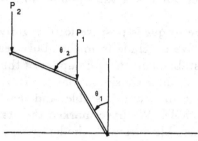

Figure 3

We again neglect the gravitational forces on the rods. The potential energy is

$$V(\theta_1, \theta_2; P_1, P_2) = \frac{1}{2}\theta_1^2 - \frac{1}{3\pi}\theta_1^3 + \frac{1}{2}(\theta_2 - \theta_1)^2 - \frac{1}{3\pi}(\theta_2 - \theta_1)^3$$
$$+ P_1 \cos\theta_1 + P_2[\cos\theta_1 + \cos\theta_2].$$

Equilibrium occurs when the partial derivatives of this function with respect to θ_1 and with respect to θ_2 are zero. It is easily seen that $\theta_1 = \theta_2 = 0$ is such an equilibrium for all values of P_1 and P_2. Other equilibria with this property are $(0, \pi)$, $(-\pi, 0)$, and (π, π). These are the only solutions when (P_1, P_2) is near zero. As before, other equilibria, which move as the vector $\mathbf{P} = (P_1, P_2)$ changes, occur. The set of equilibria can be thought of as a two-dimensional subset of the four-dimensional $(\theta_1, \theta_2, P_1, P_2)$ space. At most equilibrium points $(\theta_1, \theta_2, P_1, P_2)$ one can represent the set of equilibria in a neighborhood by prescribing θ_1 and θ_2 as smooth functions of P_1 and P_2. The points where this is not true are the bifurcation points.

As before, an equilibrium is stable if V has a local minimum there and unstable if it does not. A sufficient condition for stability is that the quadratic Taylor approximation to V have a strict local minimum. That is, an equilibrium is stable if the quadratic form $\sum \partial^2 V/\partial\theta_i\partial\theta_j \, \xi_i\xi_j$ is positive for all nonzero vectors (ξ_1, ξ_2). In this case we say that the equilibrium is **linearly stable**. If, on the other hand, there is a vector for which this sum is negative, then V does not have a local minimum at the point, so that the corresponding equilibrium is unstable. In such a situation we say that the equilibrium is **linearly unstable**.

For the equilibrium point $(0,0)$ we find the matrix

$$\frac{\partial^2 V}{\partial\theta_i\partial\theta_j}(0,0) = \begin{pmatrix} 2 - P_1 - P_2 & -1 \\ -1 & 1 - P_2 \end{pmatrix}. \tag{1.2}$$

By completing the square, we can write the corresponding quadratic form as

$$\sum_{i,j=1}^{2} \frac{\partial^2 V}{\partial\theta_i\partial\theta_j}(0,0)\xi_1\xi_j = (2 - P_1 - P_2)\left[\xi_1 - \frac{1}{2 - P_1 - P_2}\xi_2\right]^2$$
$$+ \frac{(2 - P_1 - P_2)(1 - P_2) - 1}{2 - P_1 - P_2}\xi_2^2.$$

It is clear that if the upper left entry $2 - P_1 - P_2$ and the determinant $(2 - P_1 - P_2)(1 - P_2) - 1$ of the matrix are positive, this quadratic form is positive for all nonzero ξ. If the first of these is zero or negative, then the quadratic form with $\xi_1 = 1$ and $\xi_2 = 0$ has the same property. If the upper left entry is positive but the determinant vanishes or is negative, the above identity gives the same property for the quadratic form with $\xi_1 = 1$ and $\xi_2 = 2 - P_1 - P_2$. Thus linear stability is equivalent to the two inequalities

$$2 - P_1 - P_2 > 0$$
$$(2 - P_1 - P_2)(1 - P_2) - 1 > 0.$$

The first of these is satisfied by the points of the (P_1, P_2)-plane which lie below a line. The second is satisfied by the points above the upper branch and those below the lower branch of a hyperbola one of whose asymptotes is this line. We conclude that the set of (P_1, P_2) where the solution $\theta_1 = \theta_2 = 0$ is linearly stable and hence stable consists of the region below the lower branch of the hyperbola:

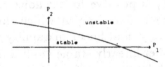

Figure 4

Above the curve the equilibrium is linearly unstable.

This diagram is called a **stability diagram** for the branch of solutions $\theta_1 = \theta_2 = 0$. As the load (P_1, P_2) approaches the boundary of the stability set, buckling will occur. In particular, we could fix P_1 and let P_2 increase gradually until it reaches the boundary of the stable set, or we could fix P_2 and let P_1 increase to the boundary. We could also prescribe the ratio of P_1 to P_2 and let them approach the stability boundary along a radial line.

We note that the stable set has two properties: firstly, it is convex, and secondly, every point to the left of and below a point of the stable set is also in this set. We shall show how to characterize and compute stable sets and how to obtain such properties.

2. Stability and buckling for finite-dimensional systems.

We now apply the ideas which were sketched in Section 1 to a general mechanical system with finitely many degrees of freedom. We suppose that we have a system such as a truss whose position is described by finitely many independent variables $(\theta_1, \cdots, \theta_n)$ which may or may not be angles. In addition, the system will depend upon finitely many parameters $(P_1, P_2, \cdots, P_m)$, which may be components of forces applied at various points of the system, but some of which may also be masses and lengths of parts of the system or spring constants. The system is described by a given potential energy function $V(\theta_1, \cdots, \theta_n; P_1, \cdots, P_m)$, whose partial derivatives with respect to the θ_j represent generalized net forces on the system.

The equilibrium states are those points at which all these forces vanish. That is, for each fixed set of parameters they are the solutions of the n equations

$$\frac{\partial V}{\partial \theta_j}(\theta_1, \cdots, \theta_n; P_1, \cdots, P_m) = 0 \qquad \text{for } j = 1, \cdots, n$$

in n unknowns. These equilibria are the **critical points** of V.

We are supposing here that there are some equations of motion which govern the system and which have the property that θ fixed at any critical point of V is a solution of these equations. We make the further assumption that the equations of motion do not permit the total energy $T + V$ of the system to increase in time, and that the kinetic energy T takes on its minimum value 0 whenever all the $d\theta_j/dt$ are 0. Then an equilibrium is stable if V has a local minimum there, and unstable otherwise.

We shall say that an equilibrium $\theta(\mathbf{P})$ is **linearly stable** if the quadratic form $\sum \partial^2 V/\partial\theta_i\partial\theta_j(\theta(\mathbf{P});\mathbf{P})\,\xi_i\xi_j$ is positive for all nonzero vectors ξ, and that it is **linearly unstable** if there exists a vector ξ at which this form is negative. As before, linear stability implies stability, and linear instability implies instability.

If an equilibrium is neither linearly stable nor linearly unstable, then the quadratic form of the second partial derivatives of V is never negative, but it is zero for some nonzero ξ. In this case we say that the quadratic form is **positive semidefinite**. The corresponding equilibrium is said to be **neutrally stable** in the mathematics literature, or **critical** in the engineering literature. Neutral stability implies neither stability nor instability, but it often indicates that the parameters are on the boundary between the sets of stability and instability. The stability of a neutrally stable state can sometimes, but not always, be determined by looking at derivatives of V of order higher than 2.

A quadratic form in a vector ξ is a polynomial which is homogeneous of degree two. Such a polynomial can always be written in the form $\xi \cdot Q\xi$ where Q is a symmetric matrix.

A symmetric matrix Q is said to be **positive semidefinite** if the corresponding quadratic form $\sum Q_{ij}\xi_i\xi_j$ never negative. It is said to be **positive definite** if this quadratic form is positive for all nonzero vectors ξ. One can therefore see whether a particular equilibrium with particular parameter values $\mathbf{P}$ is linearly stable or unstable by determining whether the matrix $\partial^2 V/\partial\theta_i\partial\theta_j$ is positive definite or positive semidefinite.

A good way of doing this is to find the minimum of the quadratic form $\sum Q_{ij}\xi_i\xi_j$ on the sphere $\sum \xi_j^2 = 1$. A standard argument with Lagrange multipliers shows that this minimum is equal to the smallest eigenvalue of the matrix. That is, it is the smallest value of λ for which the system of linear equations

$$\sum_{j=1}^{n} Q_{ij}\xi_j = \lambda\xi_i \qquad \text{for } i = 1, \cdots, n$$

has a nontrivial solution. Thus one way to check for positive definiteness is to find the lowest eigenvalue and see whether or not it is positive.

This idea is easily generalized by replacing the sphere $|\xi|^2 = 1$ by an ellipsoid. Such an ellipsoid is obtained by choosing any matrix R which is known to be positive definite and setting the quadratic form $\sum R_{ij}\xi_i\xi_j$ equal to one. The Lagrange multiplier argument now shows that the minimum of $\sum R_{ij}\xi_i\xi_j$ on this ellipsoid is the smallest value of λ for which the system

$$\sum_{j=1}^{n} Q_{ij}\xi_j = \lambda\sum_{j=1}^{n} R_{ij}\xi_j$$

has a nontrivial solution. The values of λ for which this equation has nontrivial solutions are called the **generalized eigenvalues** of the matrix Q **with respect to** the matrix R. (The usual eigenvalues are just the eigenvalues with respect to the identity matrix.)

We see, then, that Q is positive definite if and only if all its eigenvalues with respect to any particular positive definite matrix R are positive.

An important application of this fact is the following. Suppose we have a system with only one parameter P and that the potential energy V depends linearly on P. Suppose further that we are interested in the stability of a branch of equilibria θ which does not vary with P, and that this branch is linearly stable when $P = 0$. Then the matrix of second partial derivatives of V at θ has the form $R - PQ$ with R positive definite, and R and Q independent of P. This is the case, for instance, if we set $P_1 = 1/2$ and $P_2 = P$ in the two-rod example of Section 1 which leads to the matrix (1.2).

The eigenvalues of the matrix $R - PQ$ with respect to the matrix R are easily seen to have the form $1/\mu_j$, where the μ_j are the eigenvalues of the matrix Q with respect to R. Thus, the system will be linearly stable as long at the eigenvalues μ_j of Q with respect to R and the parameter P have the property that $P\mu_j < 1$ for all j. If P is positive, this simply means that P times the largest of the μ_j is less than 1. If the largest eigenvalue is positive, the system becomes linearly unstable when P exceeds its reciprocal, so that the reciprocal of the largest eigenvalue of Q with respect to R gives the critical buckling load.

If, as is often the case, the matrix Q also happens to be positive definite, we easily see that the eigenvalues of R with respect to Q are just the reciprocals of the eigenvalues of Q with respect to R. Thus in this special case the critical buckling load is the smallest eigenvalue of R with respect to Q.

The problem of determining the critical buckling load is thus reduced to the determination of the largest or the smallest eigenvalue of a generalized eigenvalue problem.

We note that the assumption that the matrix of second partial derivatives of V at the equilibrium has the form $R - PQ$ leads to a great saving of computational effort. Instead of testing the matrix for positivity for many different values of the parameter P, one computes one eigenvalue to find the whole stability set.

3. Bifurcation in continuum mechanics.

Many problems where buckling occurs involve systems whose state cannot be described by specifying finitely many numbers. Such problems arise in continuum mechanics, in which positions are specified as scalar or vector valued functions of one or more variables. This form of description is needed for a flexible beam, a flexible sheet, a bulk solid, or a system composed of such parts. In all cases the atomic nature of matter is ignored and it is assumed that such quantities as densities and stresses can be defined at all points of a domain.

To give the flavor of such a description, we consider a continuum version of the two-rod system in Section 1. This is a thin flexible and inextensible beam. (In the particular applications we shall give, such a beam is often called a column, but the methods apply equally well to what are called beam-columns.)

We shall suppose that the beam has a state in which all its points are close to a line segment of length 1, and that it is so thin that its state can be adequately described by specifying only the new shape of this line segment. Because the beam is inextensible, the line segment is not stretched or compressed. Its shape can thus be described by giving the angle $\theta(s)$ between the vertical direction and the point at distance s from the lower end.

We suppose that the lower end of the beam is firmly attached to the ground in such a way that it is always vertical. That is,

$$\theta(0) = 0. \tag{3.1}$$

As in the first example of Section 1, we shall suppose that a vertical force P pushes downward on the top end of the beam.

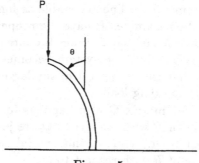

Figure 5

The contribution of this force to the potential energy is P times the height of the rod. Simple geometric considerations show that this height is equal to the integral of the function $\cos \theta(s)$ over the beam.

In addition, we assume that the beam resists bending from its straight position. The contribution per unit length of this resistance to the potential energy is a function $\psi(s, d\theta/ds)$ which describes some physical properties of the beam. The fact that the beam resists bending from the straight position means that for each s the function ψ must attain at least a local minimum at $d\theta/ds = 0$.

We shall use the notation θ' for $d\theta/ds$, with similar notation for higher derivatives.

The potential energy of a shape $\theta(s)$ is now defined as

$$V[\theta; P] := \int_0^1 [\psi(s, \theta'(s)) + P \cos \theta(s)]ds. \tag{3.2}$$

For a fixed P this formula assigns to each continuously differentiable function $\theta(s)$ with $\theta(0) = 0$ a number V. Such a function whose domain is a set of functions is called a **functional**.

We note that in place of the finitely many numbers θ_j in Section 2 we now have the infinitely many numbers $\theta(s)$. We observe that functions can be added and multiplied by numbers just like vectors. We can, in fact, think of an n-vector as a function on the integers $1, \cdots, n$. This identification of vectors with functions is a useful concept. We shall say that a set of objects, which may be vectors, functions, or vector-valued functions is a (real) **vector space** if for every pair of members ξ_1 and ξ_2 there is a member $\xi_1 + \xi_2$ of the set, and for every real number α and every member ξ of the set there is a member $\alpha\xi$, and that these operations obey the usual laws of arithmetic. These laws include the commutative law, $\xi_1 + \xi_2 = \xi_2 + \xi_1$, the associative laws $(\xi_1 + \xi_2) + \xi_3 = \xi_1 + (\xi_2 + \xi_3)$ and $\alpha(\beta\xi) = (\alpha\beta)\xi$, and the two distributive laws $\alpha(\xi_1 + \xi_2) = \alpha\xi_1 + \alpha\xi_2$ and $(\alpha_1 + \alpha_2)\xi = \alpha_1\xi + \alpha_2\xi$. One also supposes that there is a unique element 0 such that $\xi + 0 = \xi$ for all ξ, and that the number 0 times any ξ gives this element 0.

A simple example of a vector space is the set of all twice continuously differentiable functions on the interval $[0,1]$ which are zero at $s = 0$. The description of a problem includes a choice of a vector space of admissible functions.

We shall treat a system whose state is specified as a member of a given vector space S of admissible functions. The system also depends on a parameter $\mathbf{P}$, which is a member of another vector space. The system is described by a transformation $V[\theta; \mathbf{P}]$ which is, for each parameter value $\mathbf{P}$ a transformation from the space of admissible functions to the real numbers. That is, V is a family of functionals.

We assume that the potential energy functional V has the property that for any three admissible functions θ, ξ, and η the function $V[\theta + \alpha\xi + \beta\eta]$ is a twice continuously differentiable function of the pair of real variables (α, β). We define the new functional

$$\delta V[\theta; P; \xi] := \left. \frac{d}{d\epsilon} V[\theta + \epsilon\xi; P] \right|_{\epsilon=0}. \tag{3.3}$$

This functional, considered as a functional of ξ for fixed θ and P, is called the **first variation** (or Gâteau derivative) of V.

It is easily seen that the first variation is linear in ξ. By this we mean that if one replaces the function $\xi(s)$ by any linear combination $\alpha_1\xi_1(s) + \alpha_2\xi_2(s)$ of two admissible functions, the result is α_1 times the variation at ξ_1 plus α_2 times the variation at ξ_2

For the thin-beam potential functional defined in (3.2) we find that

$$\delta V[\theta; P; \xi] = \int_0^1 \left[\frac{\partial \psi}{\partial \theta'}(s, \theta'(s))\xi'(s) - P\sin(\theta(s))\xi(s) \right] ds. \tag{3.4}$$

An admissible function θ with the property that $\delta V[\theta; \mathbf{P}; \xi] = 0$ for all admissible ξ is called an **equilibrium** for the potential function at $\mathbf{P}$. In the case of the finite-dimensional potential of Section 2, this condition states that the gradient of V is zero, which reduces to the condition given there. In the case of the thin beam, we integrate the right-hand side of (3.4) by parts and use the fact that $\xi(0) = 0$ to see that

$$\delta V[\theta; P; \xi] = \int_0^1 \left[\frac{\partial \psi}{\partial \theta'}(s, \theta'(s)) - P \int_s^1 \sin(\theta(\sigma))d\sigma \right] \xi'(s)ds. \tag{3.5}$$

Since there are no constraints on the function ξ', we conclude that θ is an equilibrium if and only if it satisfies the conditions

$$\frac{\partial \psi}{\partial \theta'}(s, \theta'(s)) - P \int_s^1 \sin(\theta(\sigma))d\sigma = 0. \tag{3.6}$$

Because the coefficient of P is the indefinite integral of a smooth function, it is differentiable. Therefore we may differentiate this integrodifferential equation to obtain the differential equation

$$\frac{d}{ds}\left[\frac{\partial \psi}{\partial \theta'}(s, \theta'(s)) \right] + P\sin(\theta(\sigma(s))) = 0. \tag{3.7}$$

We need two boundary conditions for this second order equation. The first of these is (3.1). The second is obtained by setting $s = 1$ in (3.6). Thus we have the boundary conditions

$$\theta(0) = 0, \qquad \frac{\partial \psi}{\partial \theta'}(1, \theta'(1)) = 0.$$

We shall not discuss the question of how to find equilibria by solving the differential equation (3.7) with these boundary conditions. We only observe that because ψ has a minimum at $\theta' = 0$ for each fixed s, the function $\theta \equiv 0$ is an equilibrium.

In order to investigate the stability of an equilibrium, we shall use the **second variation** of V, which we define by

$$\delta^2 V[\theta; P; \xi] := \frac{d^2}{d\epsilon^2} V[\theta + \epsilon\xi; P]\bigg|_{\epsilon=0}.$$

This functional is easily shown to have the property that if $\xi(s)$ is replaced by any linear combination $\alpha_1\xi_1(s) + \alpha_2\xi_2(s) + \alpha_3\xi_3$ of three admissible functions, then the value of $\delta^2 V$ is a homogeneous polynomial of degree 2 in α_1, α_2, and α_3. We call a functional with this property a **quadratic functional**. The second variation is thus a quadratic functional.

When the vector space S consists of finite-dimensional vectors, the second variation of V reduces to the quadratic form of the matrix of its second partial derivatives.

We note that if the unknowns are functions of one or more spatial variables, the motion of the system whose equilibria we are investigating will also involve the time, so that the equation of motion is a partial differential equation. One must then be somewhat careful with statements about stability.

We still say that a quadratic functional $R[\xi]$ is positive semidefinite if it never has a negative value, and positive definite if its value is positive for all admissible ξ other than 0. However, the distinction between these two concepts is blurred in an infinite-dimensional vector space. One needs some extra machinery. We present one such sufficient condition.

STABILITY THEOREM. *Suppose that $R[\xi]$ is a quadratic functional on the vector space of admissible functions, and that it has the following properties with respect to a family $\theta = \theta(P)$ of equilibria:*

(i) *R is positive definite;*

(ii) *for the equation of motion derived by replacing the potential functional $V[\theta; P]$ by $R[\theta - \theta(P)]$, the equilibrium $\theta = \theta(P)$ is stable;*

(iii) *the ratio $\{V[\theta; P] - V[\theta(P); P] - \frac{1}{2}\delta^2 V[\theta(P); P; \theta - \theta(P)]\}/R[\theta - \theta(P)]$ approaches zero as $R[\theta - \theta(P)]$ goes to zero;*

(iv)

$$\inf_{\substack{\xi \text{ admissible} \\ \xi \neq 0}} \frac{\delta^2 V[\theta(P); P; \xi]}{R[\xi]} > 0. \tag{3.8}$$

Then the equilibrium $\theta(P)$ is stable.

If the infimum in (3.8) is negative, the equilibrium is unstable.

If the condition (3.8) holds, we shall say that the second variation of V is uniformly positive definite with respect to R, and that the equilibrium $\theta(P)$ is **linearly stable**. If the infimum of the ratio in (3.8) is negative, we say that the equilibrium is **linearly unstable**. In the case of finitely many degrees of freedom, these definitions agree with those in Section 2.

In practice, the quadratic functional R will often be the the second variation $\delta^2 V[\theta(P^{(0)}); P^{(0)}; \xi]$ at some particular parameter point $P^{(0)}$.

For the beam problem (3.2) we find that

$$\delta^2 V[\theta; P; \xi] = \int_0^1 \left[\frac{\partial^2 \psi}{\partial \theta'^2}(s, \theta'(s)) \, \xi'(s)^2 - P \cos(\theta(s) \, \xi(s)^2 \right] ds.$$

In particular, if we define the function

$$q(s) := \frac{\partial^2 \psi}{\partial \theta'^2}(s; 0),$$

then for the equilibrium solution $\theta \equiv 0$ we find that

$$\delta^2 V[0; P; \xi] = \int_0^1 \left[q(s)\xi'(s)^2 - P\xi(s)^2 \right] ds. \tag{3.9}$$

The function q represents the resistance to bending from a straight position. It is called the bending stiffness. Since ψ takes its maximum for fixed s at $\theta' = 0$, the function q must be nonnegative. We make the stronger assumption that it is uniformly positive.

It is not difficult to see that for any reasonable equation of motion the system is stable when $P = 0$, and that the quadratic functional

$$R[\xi] := \int_0^1 q(s)\xi'(s)^2 ds$$

satisfies the conditions of the Stability Theorem.

As in the discussion of Section 3, we see that the equilibrium $\theta = 0$ is linearly stable if and only if P is less than the reciprocal of the quantity

$$R^* := \sup_{\substack{\xi(0)=0 \\ \xi \not\equiv 0}} \frac{\int_0^1 \xi(s)^2 ds}{\int_0^1 q(s)\xi'^2 ds}, \tag{3.10}$$

and linearly unstable if $P > 1/R^*$. Therefore $P^* = 1/R^*$ is the critical buckling load.

By using some arguments which we shall sketch in Section 8 and which use the fact that the denominator of the ratio contains higher derivatives than the numerator, we can show that there is an admissible function $\xi^*(s)$ such that the ratio on the right of (3.10) at $\xi = \xi^*$ is equal to R^*.

Once it is known that there is such a maximizer ξ^*, we see by setting $\xi = \xi^* + \epsilon\eta$ for any admissible function η that the first variation of the ratio on the right of (3.10) at the maximizer must vanish. This gives the equation

$$\int_0^1 [\xi^*(s)\eta(s) - R^* p(s)\xi^{*\prime}(s)\eta'(s)]ds = 0$$

for all admissible functions η. A copy of the derivation of (3.7) now shows that the maximizer must solve the linear eigenvalue problem

$$[q\xi^{*'}]' + P^*\xi^* = 0$$
$$\xi^*(0) = 0, \ \xi^{*'}(1) = 0,$$

(3.11)

where $P^* = 1/R^*$.

We end this section by noting that not all equilibrium equations for mechanical systems can be derived from potentials. A simple example of this phenomenon is a two-rod system like that in Figure 3, but with the force P_2 applied in the direction of the axis of the upper rod, rather than in the vertical direction. (Such a force is often called a follower force.) The equilibrium equations for this system are

$$\theta_1 - \frac{1}{\pi}\theta_1^2 - (\theta_2 - \theta_1) + \frac{1}{\pi}(\theta_2 - \theta_1)^2 - P_1 \sin\theta_1 + P_2 \sin(\theta_2 - \theta_1) = 0$$

$$\theta_2 - \theta_1 - \frac{1}{\pi}(\theta_2 - \theta_1)^2 = 0.$$

The configuration $\theta_1 = \theta_2 = 0$ is an equilibrium, and the linearization about it is a system with the matrix $A = \begin{pmatrix} 2 - P_1 - P_2 & -1 + P_2 \\ -1 & 1 \end{pmatrix}$. The fact that this matrix is not symmetric when $P_2 \neq 0$ shows that the equilibrium equations cannot be derived from a potential function.

We see from the equilibrium equations that if we fix the value $P_1 = 0$ so that the only force is the follower force, then the only equilibria are $\theta_1 = \theta_2 = 0$ and $\theta_1 = \theta_2 = \pi$. Thus no bifurcation occurs as θ is increased. Nevertheless, one finds that for any reasonable equation of motion, the equilibrium $\theta_1 = \theta_2 = 0$ changes from being stable to being unstable as P_2 increases above a certain critical value. This critical value depends upon the masses of the rods and upon any frictional forces. Thus it cannot be predicted from the equilibrium conditions alone.

We shall not treat problems which do not come from potentials in these lectures.

4. Constraints.

It sometimes happens that a system whose state is described by a function θ, which may be a vector, a scalar-valued function, or a vector-valued function, is subjected to one or more constraints. This means that θ is not an arbitrary element of the space S, but is required to satisfy the constraint

$$K[\theta; \mathbf{P}] = 0,$$

where for each value of the parameter $\mathbf{P}$, K is an, in general nonlinear, transformation from the space of admissible functions to some vector space. If the range of K is the ℓ-dimensional vector space R^ℓ, we have ℓ constraints, but the range of K may be infinite-dimensional, in which case there are infinitely many constraints.

For example, in the two-rod system of Section 1, we may constrain the horizontal coordinate of the top of the upper rod to have a prescribed value P_2. This constraint can be written as $\sin\theta_1 + \sin\theta_2 = P_2$. Similarly, we might prescribe prescribe the horizontal coordinate P_2 of the top of the beam in Section 3, which would give the condition $\int \sin\theta(s)ds = P_2$. If one were dealing with a plate, one might want to prescribe displacements all over the boundary, which would give infinitely many constraints.

We shall assume that the constraints are independent in the sense that there is no linear functional γ on the range of K, other than the zero functional, with the property that $\gamma[\delta K[\theta; \mathbf{P}; \xi]] = 0$ for all admissible ξ, where $\delta K[\theta; \mathbf{P}; \xi]$ is the first variation of K. In the case in which θ is a finite-dimensional vector and the range of K is finite-dimensional, this simply states that the gradients of the constraint components are linearly independent.

We can then define the equilibria by the method of Lagrange multipliers. We say that θ is an equilibrium of the constrained system if and only if there is a linear functional γ on the range of K such that the system

$$\delta V[\theta; \mathbf{P}; \xi] + \gamma[\delta K[\theta; \mathbf{P}; \xi]] = 0 \qquad \text{for all admissible } \xi$$
$$K[\theta; P] = 0. \tag{4.1}$$

is satisfied. Note that when the range of K is an ℓ-space, then γ is an ℓ-vector, whose components are the usual Lagrange multipliers.

There is an immediate extension of the Stability Theorem.

CONSTRAINED STABILITY THEOREM. *Suppose that for some quadratic functional $R[\xi]$ the first three hypotheses of the Stability Theorem in Section 3 are satisfied by this R and the potential function $V + \gamma K$, where γ is the linear functional in the equilibrium condition (4.1). Then if*

$$\inf_{\substack{\xi \in S \\ \xi \neq 0 \\ \delta K[\theta; \mathbf{P}; \xi]=0}} \frac{\delta^2 V[\theta; \mathbf{P}; \xi] + \gamma \delta^2 K[\theta; \mathbf{P}; \xi]}{R[\xi]} > 0, \tag{4.2}$$

the equilibrium θ is stable.

If, on the other hand the infimum is negative, the equilibrium is unstable.

We shall again say that the equilibrium is linearly stable if (4.2) is satisfied, and that an equilibrium is unstable if the infimum in (4.2) is negative.

We shall obtain useful information about stability from the following fact, which is so obvious that it does not need a proof.

FIRST MONOTONICITY PRINCIPLE. *Let $f[\xi]$ be a functional defined for all ξ in some set T, and let T_1 be a subset of T. Then*

$$\inf_{\xi \in T_1} f[\xi] \geq \inf_{\xi \in T} f[\xi]$$

and

$$\sup_{\xi \in T_1} f[\xi] \leq \sup_{\xi \in T} f[\xi].$$

We observe that if θ is an equilibrium of the unconstrained problem which happens to satisfy the constraints $K[\theta; \mathbf{P}] = 0$, then it also satisfies the equilibrium conditions (4.1) with $\gamma = 0$. Therefore it is an equilibrium of the constrained problem. Since $\gamma = 0$, the quadratic form which appears in the stability and instability criteria is the same in the unconstrained and constrained problems. The only difference between the infimum in (3.8) for unconstrained stability and the infimum in (4.2) is that the former is taken over all nonzero members ξ of the vector space of admissible functions, while the latter is confined to the subset of those ξ which satisfy the linearized constraints $\delta K[\xi] = 0$. The First Monotonicity Principle thus has the following corollary.

COROLLARY. *If an equilibrium θ of an unconstrained problem satisfies the constraints of a constrained problem, and if it is linearly stable for the unconstrained problem, then it is also linearly stable for the constrained problem. If such a solution is linearly unstable for the constrained problem, it is also linearly unstable for the unconstrained problem.*

In other words, *imposing constraints tends to stabilize an equilibrium.* To put it another way, constraints tend to delay bifurcation.

Constraints often arise as end conditions in continuum problems.

EXAMPLE 4.1. Suppose that the top of the beam in the example of Section 3 is constrained to lie directly above (or below) the bottom. This constraint can be written as

$$\int_0^1 \sin(\theta(s))ds = 0.$$

We note that the equilibrium $\theta \equiv 0$ of the unconstrained problem satisfies this constraint, so that it an equilibrium for the constrained problem. By choosing for R the second variation (3.9) with $P = 0$, we find that the critical buckling load for the constrained problem is the reciprocal of the quantity

$$\hat{R} := \sup_{\substack{\xi(0)=0 \\ \int_0^1 \xi(s)ds=0}} \frac{\int_0^1 \xi(s)^2 ds}{\int_0^1 q(s)\xi'(s)^2 ds}. \tag{4.3}$$

Because a smaller set of ξ is admissible, the supremum in (4.3) is no larger than that in (3.10), so that, in accordance with the First Monotonicity Principle, imposing the constraint does not decrease (and usually increases) the buckling load.

One can again show by the methods to be sketched in Section 8 that there is a function $\hat{\xi}$ which satisfies the conditions in (4.3) and for which the ratio attains its maximum value $\hat{R}$. By setting $\xi = \hat{\xi} + \epsilon\eta$ and noting that the ratio in (4.3) takes on its maximum at $\epsilon = 0$ one finds that $\hat{\xi}(s)$ must satisfy the condition

$$\int_0^1 [\hat{\xi}\eta - \hat{R}q(s)\hat{\xi}'\eta']ds = 0$$

for all η which vanish at 0 and have the integral 0 over the interval [0,1]. We integrate the first term by parts to find that

$$\int_0^1 \left[\int_s^1 \hat{\xi}(\sigma)d\sigma - \hat{R}q(s)\hat{\xi}'(s)\right]\eta'(s)ds = 0 \qquad (4.4)$$

for all such η.

Because $\eta(0) = 0$, the fact that the integral of η is zero is equivalent to the condition $\int_0^1 (1-s)\eta'(s)ds = 0$. Therefore (3.12) is satisfied for all such η if and only if there is a constant α such that

$$-\hat{R}q(s)\hat{\xi}'(s) + \int_s^1 \hat{\xi}(\sigma)d\sigma = \alpha(1-s). \qquad (4.5)$$

By differentiating this equation twice we obtain a fourth order differential equation for $\hat{\xi}$. However, one of the boundary conditions is replaced by the integral condition on ξ. In order to avoid this unusual situation, we introduce the new variable

$$\hat{x}(s) := \int_0^s \hat{\xi}(\sigma)d\sigma,$$

so that $\hat{\xi} = \hat{x}'$.

If we make this change of variable after differentiating (4.5) twice, we find the differential equation

$$[q\hat{x}'']'' + \frac{1}{\hat{R}}\hat{x}'' = 0 \qquad (4.6)$$

By using the definition of $\hat{x}$ and the conditions on $\hat{\xi}$, and by setting $s = 1$ in (4.5), we find that $\hat{x}$ satisfies the boundary conditions

$$\hat{x}(0) = \hat{x}'(0) = \hat{x}(1) = \hat{x}''(1) = 0. \qquad (4.7)$$

Thus the critical buckling load for the constrained problem is the lowest eigenvalue $1/\hat{R}$ of the problem (4.6), (4.7).

If the further constraint $\theta(1) = 0$, which states that the upper end of the beam is constrained to remain vertical, is introduced, the buckling load is obtained from the same differential equation, but the boundary condition $\hat{x}''(1) = 0$ in (4.7) is replaced by the condition $\hat{x}'(1) = 0$.

The eigenvalue problem (3.11) can also be turned into a problem with the differential equation (4.6) for the function $x^*(s) := \int_0^s \xi^* d\sigma$. The boundary conditions become $x^*(0) = x^{*\prime}(0) = x^{*\prime\prime}(1) = [qx^{*\prime\prime}]'(1) + P^* x^{*\prime}(1) = 0$.

Note that the number of boundary conditions is the same for the constrained and unconstrained problems. However, imposing a constraint replaces a condition which involves a derivative of order less than half of the order 4 of the differential equation by a condition which involves a derivative of at least this order.

As a second example we consider the relaxation of a constraint.

EXAMPLE 4.2. We recall that the supremum problem (3.10) for the critical buckling load of a beam uses the condition $\xi(0) = 0$, which comes from the fact that the bottom of the beam is forced to remain vertical. Suppose that, instead, the bottom of the beam is attached to a linear spring with spring constant k which tries to keep the bottom vertical. In this case, the constraint $\theta(0) = 0$ is removed, but the term $k\theta(0)^2/2$ is added to the potential energy.

The state $\theta \equiv 0$ is still an equilibrium. By proceeding as before, we find that the reciprocal $R^{(k)}$ of the critical buckling load for this case is given by

$$R^{(k)} = \sup_{\xi \neq 0} \frac{\int_0^1 \xi(s)^2 ds}{\int_0^1 q(s)\xi'^2 ds + k\xi(0)^2}. \tag{4.8}$$

Since the linear constraint $\xi(0) = 0$ takes the problem (4.8) into (3.1), we conclude from the First Monotonicity Principle that replacing the spring by the clamping condition $\theta(0) = 0$ increases the critical buckling load.

We remark that it can again be shown that the problem (4.8) has a maximizer ξ^*. The critical buckling load is the lowest eigenvalue of a problem like (3.11), but with the boundary condition $\xi^*(0) = 0$ replaced by the condition $-\xi^{*\prime}(0) + k\xi^*(0) = 0$. We have thus shown that replacing the former condition by the latter lowers the eigenvalue.

5. The Rayleigh-Ritz method and the finite element method.

The First Monotonicity Principle provides a basis for computing an upper bound for the critical buckling load. Let $\tau_1, \tau_2, \cdots, \tau_p$ be a finite set of admissible functions which also satisfy any imposed linearized constraints in the infimum problem (4.2). The set of all linear combinations of the τ_j is a linear subspace of the space of admissible functions which satisfy the conditions in (4.2). Therefore, if the infimum of $\delta^2 V[\theta(\mathbf{P}); \mathbf{P}; \sum_{\nu=1}^p c_\nu \tau_\nu] + \gamma \delta^2 K[\theta(\mathbf{P}); \mathbf{P}; \sum_{\nu=1}^p c_\nu \tau_\nu]$ is negative, we can conclude that the infimum in (4.2) is negative, so that the equilibrium $\theta(\mathbf{P})$ is linearly unstable, and hence unstable.

We recall that the second variation of a functional V is a quadratic functional of its last variable ξ. If $Q[\xi]$ is any quadratic functional, then by definition $Q[\sum_{\nu=1}^p c_\nu \tau_\nu]$ is a homogeneous polynomial of degree two in the c_ν. By choosing one of the c_ν to be 1 and the others to be 0, we see that the coefficient of c_ν^2 must be $Q[\tau_\nu]$. By choosing all but two of the c_ν equal to 0, we see that the coefficient of $c_\mu c_\nu$ only depends on τ_μ

and τ_ν. We denote this coefficient by $2Q[\tau_\mu, \tau_\nu]$. It is easily seen that this functional of two variables is symmetric in the sense that $Q[\tau_\mu, \tau_\nu] = Q[\tau_\nu, \tau_\mu]$, and that it is linear in each of its two variables. We say that $Q[\xi, \eta]$ is a **symmetric bilinear functional**.

By choosing $\tau_\mu = \tau_\nu = \xi$ and the coefficients of all the other τ equal to zero, we see that

$$Q[\xi, \xi] = Q[\xi].$$

Thus the symmetric bilinear functional $Q[\xi, \eta]$ is determined by the quadratic functional $Q[\xi]$, and conversely.

We use these observations to write

$$\delta^2 V[\theta(\mathbf{P}); \mathbf{P}; \sum_{\nu=1}^p c_\nu \tau_\nu] + \gamma \delta^2 K[\theta(\mathbf{P}); \mathbf{P}; \sum_{\nu=1}^p c_\nu \tau_\nu]$$

$$= \sum_{\mu,\nu=1}^p \{\delta^2 V[\theta(\mathbf{P}); \mathbf{P}; \tau_\mu, \tau_\nu] + \gamma \delta^2 K[\theta(\mathbf{P}); \mathbf{P}; \tau_\mu, \tau_\nu]\} c_\mu c_\nu.$$

Thus the problem of obtaining a sufficient condition for the instability of the equilibrium is reduced to determining whether the quadratic form of a matrix takes on negative values. This problem can be solved by seeing whether the lowest eigenvalue of the symmetric matrix $\delta^2 V[\theta(\mathbf{P}); \mathbf{P}; \tau_\mu, \tau_\nu] + \gamma \delta^2 K[\theta(\mathbf{P}); \mathbf{P}; \tau_\mu, \tau_\nu]$ with respect to any positive definite matrix such as the identity is negative.

The same information can also be found by applying Gauss elimination without multiplying rows through or pivoting. If this process results in a negative leading element on the principal diagonal or a zero element without a whole row of zeros, then the equilibrium is unstable.

In the special case in which $\delta^2 V$ is linear in the single parameter P and has the form $R[\xi] - PQ[\xi]$ where both R and Q are positive definite quadratic forms, this process shows that the lowest eigenvalue of the matrix $R[\tau_\mu, \tau_\nu]$ with respect to the matrix $Q[\tau_\mu, \tau_\nu]$ is an upper bound for the critical buckling load. This idea is often called the **Rayleigh-Ritz method**.

For the beam problem of Section 3, one chooses continuously differentiable functions $\tau_\mu(s)$ which vanish at $s = 0$, and computes the lowest eigenvalue of the matrix $\int_0^1 q(s)\tau_\mu'(s)\tau_\nu'(s)ds$ with respect to the matrix $\int_0^1 \tau_\mu \tau_\nu ds$ to obtain an upper bound for the critical buckling load. The same procedure works for the constrained beam problem of Example 4.1, provided the τ_ν are chosen to satisfy the linearized constrained $\int \tau_\nu ds = 0$.

If many τ_μ are used so that the matrices are large, there is a considerable saving of computational effort if the matrices are sparse. This means that each row and each column of the matrices only contains a small number of nonzero entries.

In most problems this situation can be brought about by choosing the functions so that each τ_ν vanishes outside some small set I_ν, and that no point lies in more than a small number of these intervals. A method based on this idea is called a **finite element method**.

6. Monotonicity and convexity properties of the stability set.

In this Section we shall obtain information about the stability set by using two simple but useful facts. The first is too obvious to require a proof.

SECOND MONOTONICITY PRINCIPLE. *If the two functionals $f[\xi]$ and $g[\xi]$ have the property that $f[\xi] \leq g[\xi]$ for all ξ in some set T, then*

$$\inf_{\xi \in T} f[\xi] \leq \inf_{\xi \in T} g[\xi]$$

and

$$\sup_{\xi \in T} f[\xi] \leq \sup_{\xi \in T} g[\xi]. \tag{6.1}$$

We note the following immediate corollary of this principle.

COROLLARY: *Let $\theta = \theta(\mathbf{P})$ be a family of unconstrained equilibria for the potential energy functional $V[\theta; \mathbf{P}]$. Suppose that two values $\mathbf{P}^{(1)}$ and $\mathbf{P}^{(2)}$ of the parameter vector have the property that the inequality $\delta^2 V[\theta(\mathbf{P}^{(1)}); \mathbf{P}^{(1)}; \xi] \leq \delta^2 V[\theta(\mathbf{P}^{(2)}); \mathbf{P}^{(2)}; \xi]$ is satisfied for each admissible ξ. Then if $\theta(\mathbf{P}^{(1)})$ is linearly stable, so is $\theta(\mathbf{P}^{(2)})$. If, on the other hand, $\theta(\mathbf{P}^{(2)})$ is linearly unstable, so is $\theta(\mathbf{P}^{(1)})$.*

Roughly speaking, increasing $\delta^2 V$ makes the system more stable. For example, we see from the matrix (1.2), which is the coefficient matrix of the second variation of the potential energy for the two-rod problem, that increasing the loads P_1 and P_2 destabilizes the system. It follows that if P_2 is considered as a parameter, then the critical buckling load P_1 is a nonincreasing function of P_2, as shown in Figure 4.

In fact, we see that the quadratic form of the matrix (1.2) is nonincreasing in P_2 and in the combination $P_1 + P_2$, so that $P_1 + P_2$, with P_1 the critical buckling load corresponding to P_2, is still nonincreasing in P_2.

The matrix (1.2) comes from springs with the spring constant 1 at zero torque. The Second Monotonicity Principle shows that if we replace these springs with stronger springs (that is, springs with larger spring constants at zero torque), then the system becomes more stable. We could, of course, consider these spring constants as two more parameters and obtain this fact from the Corollary.

The same ideas can be applied to continuous problems. For example, we see from the quadratic form (3.9) for the beam problem and the Second Monotonicity Principle that replacing the stiffness function $q(s)$ by a larger function makes the system more stable. We can, in fact, think of the function $q(s)$ as an infinite set of parameters, so that, like the set of admissible functions, the set of parameters becomes an infinite-dimensional vector space.

Another example to which the Second Monotonicity Principle can be applied is the thin beam whose lower end is not constrained by the condition $\theta(0) = 0$, but is attached to a linear spring with spring constant k which tries to keep this end vertical. This beam was discussed in Example 4.2. The reciprocal of the critical buckling load is the supremum of the ratio in (4.8), which is a decreasing function of k. Thus we immediately see that the increasing the spring constant k increases the critical buckling load.

We mention the important special case in which the second variation $\delta^2 V[\theta(\mathbf{P}); \mathbf{P}; \xi]$ is linear in the parameter P_1, and, for each fixed ξ, is decreasing in all the parameters. Then, as we have seen, the critical buckling load for P_1 is an eigenvalue, and this eigenvalue is a nonincreasing function of the other parameters.

We now come to a second important property of infima and suprema. A set of vectors $\mathbf{P}$ is said to be **convex** if whenever $\mathbf{P}_1$ and $\mathbf{P}_2$ are in the set, then so is the line segment which connects them.

A function $h(\mathbf{P})$ is said to be a **concave function** of the vector variable $\mathbf{P}$ on a convex domain if for any pair of points $\mathbf{P}^{(1)}$ and $\mathbf{P}^{(2)}$ in the domain the inequality

$$h((1-t)\mathbf{P}^{(1)} + t\mathbf{P}^{(2)}) \geq (1-t)h(\mathbf{P}^{(1)}) + th(\mathbf{P}^{(2)}) \tag{6.2}$$

is valid for $0 \leq t \leq 1$. A function whose negative is concave is said to convex.

CONVEXITY PRINCIPLE. *Let the family of functionals $f[\mathbf{P}; \xi]$ have the property that for each ξ in a certain set T, $Q[\mathbf{P}; \xi]$ is a concave function of $\mathbf{P}$ on a certain convex set Π. Then*

$$\inf_{\xi \in T} f[\mathbf{P}; \xi]$$

is a concave function of $\mathbf{P}$ on Π.

In particular, the subsets of Π which consist of those $\mathbf{P}$ where this infimum is positive and where this infimum is nonnegative are both convex sets.

If, on the other hand, the functional $f[\mathbf{P}; \xi]$ is convex for each ξ in T, then $\sup_{\xi \in T} f[\mathbf{P}; \xi]$ is convex.

Proof. For each nonzero ξ the function $h(\mathbf{P}) = f[\mathbf{P}; \xi]$ is concave. Hence we see from (6.2) and the definition of infimum that

$$f[(1-t)\mathbf{P}^{(1)} + t\mathbf{P}^{(2)}; \xi] \geq (1-t)f[\mathbf{P}^{(1)}; \xi] + tf[\mathbf{P}^{(2)}; \xi]$$
$$\geq (1-t)\inf_{\eta \in T} f[\mathbf{P}^{(1)}; \eta] + t\inf_{\eta \in T} f[\mathbf{P}^{(2)}; \eta]$$

when $0 \leq t \leq 1$. By taking the infimum of the left-hand side, we see that

$$\inf_{\xi \in T} f[(1-t)\mathbf{P}^{(1)} + t\mathbf{P}^{(2)}; \xi] \geq (1-t)\inf_{\eta \in T} f[\mathbf{P}^{(1)}; \eta] + t\inf_{\eta \in T} f[\mathbf{P}^{(2)}; \eta],$$

so that the infimum is concave.

It is clear from (6.2) that if h is concave and if its value is positive or nonnegative at $\mathbf{P}^{(1)}$ and $\mathbf{P}^{(2)}$, the same is true on the line segment which joins these two points. This proves the second statement of the Convexity Principle.

To prove the last statement, we only need to notice that the supremum of f is minus the infimum of $-f$, and to apply the first statement to the concave function $-f$.

The Convexity Principle can immediately be applied to the infimum (3.8) of the Stability Theorem and to the infimum (4.2) of the Constrained Stability Theorem to obtain the following corollary.

COROLLARY. *If $\delta^2 V[\theta(\mathbf{P}; \mathbf{P}; \xi]$ is convex in $\mathbf{P}$ for $\mathbf{P}$ in a convex set Π, then for the unconstrained problem the set of $\mathbf{P} \in \Pi$ for which the system is linearly stable and the complement of the set for which the system is linearly unstable are convex.*

The same is true for the constrained problem, provided one can find a convex set Π of $\mathbf{P}$ with the following properties: The set T of nonzero ξ in S which satisfy the linearized constraint $\delta K[\mathbf{P}; \xi] = 0$ is the same for all $\mathbf{P}$ in Π, and the numerator in (4.2) is concave in Π for each fixed ξ.

The simplest case again occurs when V is linear in $\mathbf{P}$ and the branch of equilibria is independent of $\mathbf{P}$, and when there are no constraints. In this case the second variation of V is linear in $\mathbf{P}$, so that it is automatically concave.

If we make the additional assumption that the the second variation of the coefficient of P_1 is the negative of a positive definite quadratic functional, the Second Monotonicity Principle and the Convexity Principle imply that the critical buckling load P_1^* is a concave function of the other parameters.

These conditions are valid in the two-rod example of Section 1. The Convexity Principle thus predicts the convexity of the set of points below the hyperbola in Figure 4, and the fact that the critical value of P_2 is a concave function of P_1.

EXAMPLE 6.1 (Southwell summation). Consider the vertical inextensible beam of height 1 shown in Figure 5, but with its top attached to a spring of length 1 fastened at the point (0,2). We now call the downward force at the top P_1. Let the bending stiffness $q(s)$ of the column be a constant P_2 and call the spring constant of the spring P_3. Then

$$V[\theta; \mathbf{P}] = \frac{1}{2} P_2 \int_0^1 \theta'^2 dz + \frac{1}{2} P_3 \left[\left\{ 2 - \int_0^1 \cos\theta \, ds \right\}^2 + \left\{ \int_0^1 \sin\theta \, ds \right\}^2 \right]$$
$$+ P_1 \int_0^1 \cos\theta \, ds.$$

We still require that $\theta(0) = 0$. The second variation about the equilibrium $\theta(s) \equiv 0$ is given by

$$\delta^2 V[0; \xi; \mathbf{P}] = P_2 \int_0^1 \xi'^2 ds + P_3 \left[\int_0^1 \xi^2 dz + \left\{ \int_0^1 \xi \, ds \right\}^2 \right] - P_1 \int_0^1 \xi^2 dz.$$

Define $P_{cr}^{(2)}$ to be the critical buckling value of the force P_1 when the spring constant P_3 is replaced by 0. (That is, the spring is removed.) Let $P_{cr}^{(3)}$ be the critical value of P_1 when the bending stiffness is replaced by 0. (That is, the beam is removed.) Then the column is neutrally stable at the values $(P_{cr}^{(2)}, P_2, 0)$ and $(P_{cr}^{(3)}, 0, P_3)$ of the vector parameter. The convexity principle states that the system is at least neutrally stable at their average $\frac{1}{2}(P_{cr}^{(2)} + P_{cr}^{(3)}, P_2, P_3)$. Because $\delta^2 V$ is linear and homogeneous in $\mathbf{P}$, multiplying a vector $\mathbf{P}$ by a positive constant does not change its stability property. Therefore, the system is also at least neutrally stable at $(P_{cr}^{(2)} + P_{cr}^{(3)}, P_2, P_3)$. This means that the critical buckling load P_{cr} of P_1 which corresponds to the given P_2 and P_3 must be at least as large as the first entry of this vector. That is,

$$P_{cr} \geq P_{cr}^{(2)} + P_{cr}^{(3)}.$$

This consequence of the convexity Principle is called a Southwell Summation Theorem. (See the article of T. Tarnai in this Volume.)

We remark that because of the homogeneity, $P_{cr}^{(2)}$ is equal to P_2 times the value which corresponds to $P_2 = 1$, $P_3 = 0$. Similarly, $P_{cr}^{(3)}$ is P_3 times its value for $P_2 = 0$, $P_3 = 1$. It is, in fact, easy to see that $P_{cr}^{(3)} = 2P_3$.

EXAMPLE 6.2 (Dunkerley theorem). We return to the system of two rods pictured in Figure 3 of Section 1. The second variation of the potential energy at $\theta_1 = \theta_2 = 0$ is the quadratic form of the matrix (1.2). It is usual to fix the ratio of the two forces by introducing the reference forces a_1 and a_2, and setting

$$P_1 = \lambda a_1, \qquad P_2 = \lambda a_2,$$

where λ is a positive parameter. If we insert these values into the quadratic form of (1.2) and divide by λ, we obtain the quadratic form

$$\frac{1}{\lambda}[2\xi_1^2 - 2\xi_1\xi_2 + \xi_2^2] - a_1\xi_1^2 - a_2[\xi_1^2 + \xi_2^2]$$

The stability of the system for a particular parameter vector $(1/\lambda, a_1, a_2)$ is determined by whether or not this form is positive definite or semidefinite.

We define the critical value $\lambda_{cr}^1(1)$ which is obtained when a_2 is replaced by zero, so that $P_2 = 0$, and the critical value $\lambda_{cr}^{(2)}$ which is obtained when a_2 is replaced by zero. We note that the above quadratic form is linear homogeneous in the components of the parameter vector. Therefore we can use the convexity principle in the same manner as in the preceding example to show that the system is at least neutrally stable at the parameter point $(1/\lambda_{cr}^{(1)} + 1/\lambda_{cr}^{(2)}, a_1, a_2)$.

We conclude that the critical value λ_{cr} for the given a_1 and a_2 must satisfy the inequality

$$\frac{1}{\lambda_{cr}} \leq \frac{1}{\lambda_{cr}^{(1)}} + \frac{1}{\lambda_{cr}^{(2)}}.$$

This consequence of the convexity principle is called a Dunkerley Summation Theorem. (See the article of T. Tarnai in this Volume.)

We can show as in the previous example that each of the $1/\lambda_{cr}^{(i)}$ is equal to a constant which depends on i times a_i.

7. Completion of a vector space.

In this Section we introduce some techniques of functional analysis which will serve two purposes. In the first place, we will show how to extend the vector space of admissible functions in a problem of the form

$$\sup_{\substack{\xi \in S \\ \xi \neq 0}} \frac{Q[\xi]}{R[\xi]}, \tag{7.1}$$

where Q and R are quadratic functionals on a prescribed vector space S of admissible functions, and R is positive definite, without increasing the value of the supremum. This will make it easier to find functions τ_ν which can be used in the Rayleigh-Ritz method.

The same process will also be used in the next Section to give some conditions which assure us that there is a maximizer ξ^* at which the ratio attains its supremum. As we have seen in Sections 3 and 4, the existence of such a maximizer often leads to the characterization of the supremum in terms of an eigenvalue of a differential operator.

We have shown in Section 5 that to any quadratic functional $R[\xi]$ one can associate a symmetric bilinear functional $R[\xi, \eta]$ with the property that if ξ and η are any two elements of S, then

$$R[\alpha\xi + \beta\eta] = \alpha^2 R[\xi] + \beta^2 R[\eta] + 2\alpha\beta R[\xi, \eta]. \tag{7.2}$$

Because we have assumed that $R[\xi]$ is positive definite, the quadratic polynomial on the right-hand side of (7.2) is never negative, and it is positive unless either $\alpha = \beta = 0$ or ξ and η are linearly dependent. This means that the quadratic equation for β obtained by setting the right-hand side of (7.2) equal to zero and choosing $\alpha = 1$ cannot have two distinct real roots. Therefore the discriminant $R[\xi, \eta]^2 - R[\xi]R[\eta]$ cannot be positive, and it is zero if and only if ξ and η are linearly dependent. We write this inequality in the form

$$|R[\xi, \eta]| \leq R[\xi, \xi]^{1/2} R[\eta, \eta]^{1/2}, \tag{7.3}$$

with equality if and only if ξ and η are linearly dependent.

We shall think of $R[\xi, \eta]$ as a scalar product between vectors, and $R[\xi, \xi]^{1/2}$ as the length of ξ. Then this is just the usual **Schwarz's inequality**, which says that one can define the angle between two vectors by saying that the cosine of this angle is the scalar product divided by the product of the lengths.

We shall call the vector space S with the scalar product $R[\xi, \eta]$ and the corresponding concept of the distance $R[\xi - \eta, \xi - \eta]^{1/2}$ between two functions an **inner product space**.

We see from (7.2) with $\alpha = \beta = 1$ and the Schwarz inequality (7.3) that

$$R[\xi + \eta]^{1/2} \leq R[\xi]^{1/2} + R[\eta]^{1/2}. \tag{7.4}$$

This is known as the **triangle inequality**, and states that the sum of the lengths of two sides of a triangle is at least equal to the length of the third side. That is, the shortest distance between two points is a straight line.

A concept of distance produces a concept of convergence. We say that an infinite sequence $\{\xi_j\}$ of members of the inner product space S converges to ξ when

$$\lim_{j \to \infty} R[\xi - \xi_j] = 0. \tag{7.5}$$

We say that a linear functional $\ell[\xi]$ is **bounded** on the inner product space S if the ratio $\ell[\xi]^2 / R[\xi, \xi]$ is uniformly bounded. We define the **norm** of ℓ as

$$||\ell|| = \left\{ \sup_{\substack{\xi \in S \\ \xi \neq 0}} \frac{\ell[\xi]^2}{R[\xi, \xi]} \right\}^{1/2}. \tag{7.6}$$

Because $\ell[\xi]^2$ is a quadratic functional, the supremum problem in (7.6) is a special case of (7.1), and we ask whether this supremum is attained.

Assume that ℓ is not the zero functional, so that $||\ell|| > 0$. By definition of a supremum, there is a sequence of elements $\{\xi_j\}$ of S with the properties $\ell[\xi_j] = ||\ell||^2$ and $\lim_{j \to \infty} R[\xi_j, \xi_j] = ||\ell||^2$.

Let η be any nonzero element of S for which

$$\ell[\eta] = 0.$$

Then for any real α we have $\ell[\xi_j + \alpha\eta] = ||\ell||^2$, so that by (7.6), $R[\xi_j + \alpha\eta] \geq ||\ell||^2$. In particular, the choice $\alpha = -R[\xi_j, \eta]/R[\eta]$ leads to the inequality

$$R[\xi_j, \eta]^2 \leq R[\eta]\{R[\xi_j] - ||\ell||^2\} \qquad \text{when } \ell[\eta] = 0. \tag{7.7}$$

We first consider the special case in which $\eta = \xi_k - \xi_j$ for some j and k. We use the square root of the resulting inequality and the same inequality with j and k interchanged to see that

$$R[\xi_k - \xi_j, \xi_k - \xi_j] = R[\xi_k, \xi_k - \xi_j] - R[\xi_j, \xi_k - \xi_j]$$

$$\leq R[\xi_k - \xi_j, \xi_k - \xi_j]^{1/2} \left(\{R[\xi_k] - ||\ell||^2\}^{1/2} + \{R[\xi_j] - ||\ell||^2\}^{1/2} \right).$$

By dividing both sides by the norm $R[\xi_k - \xi_j]^{1/2}$ and recalling that the sequence $R[\xi_j]$ converges to $||\ell||^2$, we see that

$$\lim_{j,k \to \infty} R[\xi_k - \xi_j] = 0. \qquad (7.8)$$

That is, any maximizing sequence satisfies the so-called Cauchy criterion (7.8), which simply states that the members of the sequence get close to each other.

It follows immediately from the triangle inequality that if the sequence $\{\xi_j\}$ converges to some ξ, then it satisfies the Cauchy criterion. In a finite-dimensional space the Cauchy criterion implies that the sequence converges to some element of S, but this need not be true in an infinite-dimensional space. Consider for example the space S of of continuous functions on the interval $[0,1]$ with the scalar product $R[\xi, \eta] = \int_0^1 \xi(s)\eta(s)ds$, The sequence of continuous functions

$$\xi_n = \begin{cases} 1 & \text{for } \xi \leq \frac{1}{2} \\ 1 - j\left(x - \frac{1}{2}\right) & \text{for } \frac{1}{2} \leq x \leq \frac{1}{2} + \frac{1}{j} \\ 0 & \text{for } x \geq \frac{1}{2} + \frac{1}{j} \end{cases}$$

satisfies the Cauchy criterion, but does not converge to a continuous function. It does converge to a function with a jump, so that if it is to have a limit, we must extend the space to include this function.

For a general inner product space, we need to make such an extension for every bounded linear functional for which the supremum in (7.6) is not attained by an element of S.

This is most easily done in the following manner. Define the **dual space** S' of the inner product space S to be the set of all bounded linear functionals. We observe that if ℓ and m are two bounded linear functionals and α and β are real numbers, we can define a new bounded linear functional

$$(\alpha\ell + \beta m)[\xi] := \alpha\ell[\xi] + \beta m[\xi].$$

Therefore S' is again a vector space. Moreover, since $|\ell[\xi] + m[\xi]| \leq |\ell[\xi]| + |m[\xi]|$, we see that the triangle inequality $||\ell + m|| \leq ||\ell|| + ||m||$ is satisfied, so that it is natural to think of the norm as a distance.

Any member ζ of S defines the bounded linear functional

$$\ell_\zeta[\xi] := R[\zeta, \xi].$$

The transformation from ζ in S to ℓ_ζ in S' is easily seen to be a linear transformation. We see from Schwarz's inequality (7.3) and the fact that $\ell_\zeta[\zeta] = R[\zeta]$ that $||\ell_\zeta|| = R[\zeta]^{1/2}$, so that the transformation is an isometry.

We now return to the arbitrary bounded linear functional ℓ and the maximizing sequence ξ_j for (7.6). We observe that for any function ζ in S the function $\eta =$

$\ell[\xi_j]\zeta - \ell[\zeta]\xi_j$ satisfies $\ell[\eta] = 0$. Therefore we may use the the inequality (7.7) and the triangle inequality (7.4) to see that

$$|\ell[\xi_j](\ell_{\xi_j}[\zeta] - \ell[\zeta])| = |R[\xi_j, \ell[\xi_j]\zeta - \ell[\zeta]\xi_j] + \ell[\zeta](R[\xi_j] - \ell[\xi_j])|$$

$$\leq R[\ell[\xi_j]\zeta - \ell[\zeta]\xi_j]^{1/2}\left\{R[\xi_j] - ||\ell||^2\right\}^{1/2} + |\ell[\zeta]|\left\{R[\xi_j] - ||\ell||^2\right\}$$

$$\leq R[\zeta]^{1/2}[(|\ell[\xi_j]| + ||\ell||R[\xi_j]^{1/2})\left\{R[\xi_j] - ||\ell||^2\right\}^{1/2} + ||\ell||\left\{R[\xi_j] - ||\ell||^2\right\}].$$

Because the maximizing sequence $\{\xi_j\}$ was chosen so that $\ell[\xi_j] = ||\ell||^2$, and $R[\xi_j]$ converges to $||\ell||^2$, we conclude that

$$\lim_{k\to\infty} ||\ell - \ell_{\xi_k}|| = 0. \tag{7.9}$$

We have shown that every member of S' is the limit in the norm of elements in the image of the one-to-one linear transformation $\eta \to \ell_\eta$. We say that this transformation gives an imbedding of S to a dense subspace of S'. We shall identify the function ξ with its image ℓ_ξ in S', and think of S' as an extension of S.

We define the bilinear functional $R[\ell_\xi, \ell_\eta] := R[\xi, \eta]$ on the image of the linear transformation. Because every ℓ in S' is the limit in norm of a Cauchy sequence ℓ_{ξ_j}, we can extend this definition to all of S'. If ℓ is a bounded linear functional ℓ with maximizing sequence $\{\xi_j\}$ and m has the maximizing sequence $\{\eta_k\}$,

$$R[\ell, m] := \lim_{j,k\to\infty} R[\xi_j, \eta_k] = \lim_{j\to\infty} m[\xi_j] = \lim_{k\to\infty} \ell[\eta_k].$$

It is easily checked that this functional is still symmetric and bilinear, and that $R[\ell, \ell] = ||\ell||^2$, so that the quadratic functional is still positive definite. The space S' with the scalar product $R[\ell, m]$ is again an inner product space.

However, S' has an additional property. Let $\{m_j\}$ be any sequence of bounded linear functionals and let the Cauchy criterion

$$\lim_{j,k\to\infty} R[m_j - m_k, m_j - m_k] = 0$$

be satisfied. Then for any ξ in S the sequence of numbers $\{m_j[\xi]\}$ satisfies the Cauchy criterion, and therefore has a limit, which we call $\ell[\xi]$. By letting j go to infinity in the inequality $|m_j[\xi] - m_k[\xi]| \leq R[m_j - m_k, m_j - m_k]^{1/2}R[\xi, \xi]^{1/2}$ and using the fact that the Cauchy criterion is satisfied, we find that

$$\lim_{k\to\infty} ||\ell - m_k|| = \lim_{k\to\infty} R[\ell - m_j, \ell - m_j]^{1/2} = 0.$$

In particular, we note that the limit functional ℓ is bounded so that it is again in S', and that its norm is the limit of $||m_j||$.

We have thus shown that every Cauchy sequence m_j in S' has a limit in S'. A space with this property is said to be **complete**. A complete inner product space is called a **Hilbert space**. We have thus shown how to imbed any inner product space as a dense subspace of a Hilbert space.

A Hilbert space is its own dual space. That is, every bounded linear functional can be written as the scalar product with one of its members.

We now return to the problem of finding the supremum (7.1) under the additional hypothesis that there is a constant c such that

$$|Q[\xi]| \leq cR[\xi]. \tag{7.10}$$

We have already seen how to extend the quadratic functional R to the complete space S' by continuity. We now observe that the quadratic functional $(c+1)R[\xi]-Q[\xi]$ is bounded below by $R[\xi]$, so that it is positive definite, and that it is bounded above by $(2c+1)R[\xi]$. Therefore, a Cauchy sequence with respect to the norm $R^{1/2}$ is also a Cauchy sequence with respect to the norm $\{(c+1)R - Q\}^{1/2}$. Thus if $\{\xi_j\}$ is a maximizing sequence for an ℓ in S', we find that $\lim_{j\to\infty}\{(c+1)R[\xi_j] - Q[\xi_j]\}$ exists. Since we already know that the sequence $R[\xi_j]$ has the limit $R[\ell]$, we can define

$$Q[\ell] := \lim Q[\xi_j].$$

One easily checks that this extension of Q is a quadratic functional on S'. Therefore, the ratio in (7.1) is defined on S' as the limit of its values in S. It follows that the supremum of this ratio is not changed. We summarize this fact as follows.

EXTENSION LEMMA. *If S' is the completion, that is, the dual space, of the space S with the inner product $R[\xi,\eta]$, then*

$$\sup_{\ell\in S'} \frac{Q[\ell]}{R[\ell]} = \sup_{\xi\in S} \frac{Q[\xi]}{R[\xi]}.$$

REMARK. Because an infimum of a function can be written as minus the supremum of the negative of the function, the Extension Lemma is also true for an infimum.

EXAMPLE 7.1. If S is the space of all continuous functions on the interval $[0,1]$ and $R[\xi] = \int_0^1 \xi^2 ds$, and if $\ell(s)$ is a piecewise continuous bounded function, then by Schwarz's inequality the linear functional $\ell[\xi] := \int_0^1 \ell(s)\xi(s)ds$ is bounded. By extending the definition of integration to the Lebesgue integral, we can define such a bounded linear functional even when the function $\ell(s)$ is highly discontinuous, as long as the integral of its square is finite. It can be shown that every member of the completion S' is obtained in this way. That is, we can say that the space S' is the space of all Lebesgue square integrable functions. Note that two functions which differ on a sufficiently small set of points induce the same functional. Therefore we must identify such functions as the same function before we can say that the extended form R is positive definite.

EXAMPLE 7.2. If S is the space of continuously differentiable functions in the interval $[0,1]$ which vanish at $s = 0$, and $R[\xi] := \int_0^1 q(s)\xi'(s)^2 ds$ with q positive and bounded away from zero, then one can show that every bounded linear functional has the representation

$$\ell[\xi] = \int_0^1 q(s)\hat{\ell}(s)\xi'(s)ds,$$

where the function $\hat{\ell}$ is Lebesgue square integrable. In order to extend the definition of the bilinear form R, we would like the function $\hat{\ell}$ to be the derivative of a function ℓ which vanishes at $s = 0$. Accordingly, we define

$$\ell(s) := \int_0^s \hat{\ell}(\sigma)d\sigma.$$

The completion S' of the space S can thus be thought of consisting of all functions of this form with $\hat{\ell}$ Lebesgue square integrable. Such functions are continuous, but need not be differentiable.

The Extension Lemma permits us to use this extended class of functions in the Rayleigh-Ritz method for bounding the critical load of the thin beam problem whose reciprocal is given by (3.8). For example, we may use piecewise linear functions. If we divide the interval $[0,1]$ into p equal parts and let τ_ν be the function which is 1 at the right end point of the νth interval and zero at all the other end points and which is linear on each interval, we have a set of finite elements, so that the matrices which determine the upper bound for the critical buckling load are sparse.

If $R[\xi]$ and $\hat{R}[\xi]$ are two positive definite functionals on a vector space S, and if the ratio $\hat{R}[\xi]/R[\xi]$ is bounded above and below for all $\xi \neq 0$ in S, we say that the norms $\hat{R}[\xi]^{1/2}$ and $R[\xi]^{1/2}$ are **equivalent**. We observe that a linear functional on S is bounded with respect to one of these norms if and only if it is bounded with respect to the other. Therefore the completion S' is the same with respect to either of these norms. For example, as long as the function q has positive upper and lower bounds, the completion of the continuously differentiable functions with respect to the norm $\{\int_0^1 [q\xi'^2 + \xi^2]dx\}^{1/2}$ is the same as that with respect to the norm $\{\int_0^1 [\xi'^2 + \xi^2]dx\}^{1/2}$.

For this reason it is useful to deal with certain standard spaces. In a domain D in any number n of dimensions we define the Sobolev space $H^k(D)$ to be the completion of the space $C^k(D)$ of functions with continuous partial derivatives of all orders up to k with respect to the norm

$$\|\xi\|_{H^k} := \left\{ \int \sum_{\alpha_1 + \cdots + \alpha_n \leq k} [\partial^{\alpha_1, \cdots, \alpha_n} \xi]^2 dx_1 \cdots dx_n \right\}^{1/2},$$

where the symbol $\partial^{\alpha_1, \cdots, \alpha_n} \xi$ means the partial derivative of ξ which is of order α_1 with respect to x_1, of order α_2 with respect to x_2, and so forth.

The Extension Lemma sometimes permits one to remove some of the constraints in the constrained infimum problem in (4.2) without changing the value of the supremum. The infimum in (4.2) is taken not over the space S but over the subspace S_K of elements of S which satisfy the linearized constraints $\delta K[\xi] = 0$.

Suppose, for the moment that the range of the constraint functional K is one-dimensional, so that there is only one linearized constraint $\delta K[\xi] = 0$, and that this constraint is unbounded. That is, there is a sequence $\{\xi_n\}$ of elements of S such that $\delta K[\xi_n] = 1$ for all n, while $R[\xi_n]$ converges to 0 as $n \to \infty$. Then if ξ is any element of S, the sequence $\{\xi - \delta K[\xi]\xi_n\}$ satisfies the linearized constraint, and converges to ξ in the norm $R[\eta]^{1/2}$ as $n \to \infty$.

We thus see that if the single functional δK is unbounded, the completion of the subspace S_K contains all of S, and therefore coincides with the completion of S. Thus the Extension Theorem tells us that we may ignore an unbounded linearized constraint.

The above observation permits us to ignore certain boundary conditions in constructing trial functions for the Rayleigh-Ritz method.

EXAMPLE 7.3. We saw that the problem (3.10) leads to the boundary conditions in (3.11). Consider the analogue of the problem (3.10) with the additional constraint $\xi'(1) = 0$. The sequence $\xi_n = [e^{n(s-1)} - e^{-n}]/n$ satisfies the conditions $\xi_n(0) = 0$ and $\xi_n'(1) = 1$, while $R[\xi_n] = \int_0^1 q\xi_n'^2 dx$ approaches zero. This shows that the linear functional $\xi'(1)$ is unbounded. Therefore, one obtains the same supremum with or without the condition $\xi'(1) = 0$. In particular, the functions in the Rayleigh-Ritz method do not need to satisfy this extra condition. This permits the use to a piecewise linear function which is not constant near $s = 1$.

We note that even though the boundary condition $\xi'(1) = 0$ of the boundary value problem (3.11) is ignored in the maximizing process (3.10), the maximizing function, if it exists, has this property. Such a condition is called a **natural boundary condition**.

The idea of reducing the number of constraints when at least one of them is unbounded can be extended to the case in which K consists of any finite number of constraints. In this case one can show that the completion of S_K is the same as the completion of the space which is obtained from S by imposing only those linear combinations $\gamma\delta K[\xi] = 0$ of the conditions for which the linear functionals $\gamma\delta K$ are bounded. A similar result applies to the case of infinitely many constraints, provided the operator δK satisfies some additional conditions.

Because we are only interested in finding the supremum in (7.10) The two-sided boundedness condition (7.10) is not quite natural. If we only assume that

$$Q[\xi] \leq cR[\xi],$$

we observe that finding the supremum for the problem (7.1) is equivalent to finding

$$\sup_{\substack{\xi \in S \\ \xi \neq 0}} \frac{R[\xi]}{(c+1)R[\xi] - Q[\xi]}.$$

We note that the new denominator is at least equal to $R[\xi]$, so that it is positive definite, and that the ratio in this problem lies between 0 and 1. We are thus led to complete the space S with respect to the new inner product $(c+1)R[\xi,\eta] - Q[\xi,\eta]$. Because the new length is bounded below by the old length but may not be bounded above by a multiple of this length, this process may lead to a smaller extension of S than that which would be obtained from the inner product $R[\xi,\eta]$.

8. A condition for the existence of a maximizer.

As we saw in Sections 3 and 4, the existence of a maximizer of the supremum problem (7.1) often permits one to characterize the supremum as an eigenvalue of a differential operator. Thus, it is useful to have a sufficient condition for this existence.

Since by the definition of the extension of the bilinear functional R from S to S' we have $\ell[\xi] = R[\ell,\xi]$, the Extension Lemma tells us that the supremum in (7.6) is equal to the supremum over m in S' of the ratio $R[\ell,m]^2/R[m,m]$. This supremum is $\|\ell\|^2$, and it is attained for $m = \ell$. Thus completing the space can lead to the existence of a maximizer when there is none in the original problem.

However, it is not true that any supremum is attained on a Hilbert space. Consider, for example, the ratio $\int_0^1 \xi'^2 ds / \int_0^1 (\xi'^2 + \xi^2)ds$ on the space S of continuously differentiable functions which vanish at 0. The completion S' of this space again consists of indefinite integrals of Lebesgue square integrable functions. The ratio is clearly less than 1, and the sequence $\xi_j = \sin j\pi s$ shows that, in fact, the supremum is 1. However, the only way for the ratio to be 1 on the completion S' is for the integral xi^2 to be zero, which means that the function is zero. That is, there is no maximizer on the Hilbert space S'.

We will present a sufficient condition for the maximizer of the problem (7.1) to exist on the completion S' of S.

MAXIMIZATION CONDITION. *Suppose that that there exists a finite collection of bounded linear functionals $\{\ell_1, \ell_2, \cdots, \ell_r\}$ in the dual space with the property*

$$\sup_{\substack{\xi \in S \\ \ell_1[\xi]=\cdots=\ell_r[\xi]=0 \\ \xi \neq 0}} \frac{Q[\xi]}{R[\xi]} < \sup_{\substack{\xi \in S \\ \xi \neq 0}} \frac{Q[\xi]}{R[\xi]}. \tag{8.1}$$

Then there is an element ξ^* in the completion S' of S such that the extension of the ratio Q/R to S' evaluated at ξ^* is equal to the supremum on the right of (8.1).

Proof. Let μ denote the supremum on the right of (8.1), and let the supremum of the left of (8.1) be $\mu - \sigma$. By hypothesis, $\sigma > 0$. Let $\{\xi_j\}$ be a maximizing sequence for the supremum on the right with the property that $R[\xi_j] = 1$ for all j while $Q[\xi_j]$ increases to μ. Then the sequences of numbers $\{\ell_i[\xi_j]\}, \cdots, \{\ell_r[\xi_j]\}$ are all bounded, so that there is a subsequence $\hat{\xi}_j$ for which these sequences all converge.

If the linear functionals ℓ_ν are linearly dependent, the equations $\ell_\nu[\xi] = 0$ are redundant, and we can eliminate some of the ℓ_ν which are linear combinations of the others to arrive at a linearly independent set. We suppose that this has been done to begin with. Then we can find functions $\eta_1, \cdots, \eta_r$ in S such that

$$\ell_\nu[\eta_\mu] = \begin{cases} 1 & \text{if } \mu = \nu \\ 0 & \text{if } \mu \neq \nu. \end{cases}$$

We define the new sequence

$$\zeta_j = \hat{\xi}_j - \sum_{\nu=1}^{r} \ell_\nu[\hat{\xi}_j]\eta_\nu, \tag{8.2}$$

so that $\ell_1[\zeta_j] = \cdots = \ell_r[\zeta_j] = 0$. Then by hypothesis $Q[\zeta_j - \zeta_k] \leq (\mu - \sigma)R[\zeta_j - \zeta_k]$ for all j and k. We write this inequality as

$$\sqrt{\sigma}R[\zeta_j - \zeta_k]^{1/2} \leq \{\mu R[\zeta_j - \zeta_k] - Q[\zeta_j - \zeta_k]\}^{1/2}.$$

We use the definition (8.2) of ζ_j and apply the triangle inequality (7.4) for the positive semidefinite quadratic functional $\mu R - Q$ to find that this inequality implies that

$$\sqrt{\sigma}R[\zeta_j - \zeta_k]^{1/2} \leq \{\mu R[\hat{\xi}_j] - Q[\hat{\xi}_j]\}^{1/2} + \{\mu R[\hat{\xi}_k] - Q[\hat{\xi}_k]\}^{1/2}$$
$$+ \sum_{\nu=1}^{r} |\ell_\nu[\hat{\xi}_j] - \ell_\nu[\hat{\xi}_k]|\{\mu R[\eta_\nu] - Q[\eta_\nu]\}^{1/2}.$$

Because $\{\hat{\xi}_j\}$ is a maximizing sequence for the right-hand side of (8.1), the first two terms on the right approach zero as j and k go to infinity. Because the sequences $\{\ell_\nu[\hat{\xi}_j]\}$ converge, the terms of the sum on the right also go to zero.

We conclude that the maximizing sequence $\{\hat{\xi}_j\}$ satisfies the Cauchy criterion. As we have seen in Section 7, this implies that the sequence has a limit ξ^* in the completion S'. For this ξ^* the ratio in (8.1) is equal to the supremum on he right, so the Maximization Condition is proved.

In order to apply the maximization condition, we need to be able to construct linear functionals for which its hypothesis is satisfied. The most useful tool for this purpose is the following:

POINCARÉ INEQUALITY. *Let ξ be any function which is continuously differentiable on an interval $[a, b]$. Then the condition*

$$\int_a^b \xi(s)ds = 0$$

implies that

$$\int_a^b \xi^2 ds \leq \frac{(b-a)^2}{\pi^2} \int_a^b \xi'^2 ds. \tag{8.3}$$

Proof. We first observe that if we introduce the new variable $t = (s - a)/(b - a)$, then

$$\int_a^b \xi(s)^2 \, ds = (b - a) \int_0^1 \xi(a + (b - a)t)^2 \, dt$$

and

$$\int_a^b \left\{ \frac{d}{ds} [\xi(s)] \right\}^2 ds = \frac{1}{b - a} \int_0^1 \left\{ \frac{d}{dt} [\xi(a + (b - a)t)] \right\}^2 dt.$$

therefore, it is sufficient to prove the inequality (8.3) for $a = 0$ and $b = 1$. We shall assume this from now on.

For $0 < s_1 < s_2 < 1$ we can write $\xi(s_2) - \xi(s_1)$ as the integral of ξ'. By writing this integrand as $\sqrt{\sin \pi s}(\xi'(s)/\sqrt{\sin \pi s})$ and using Schwarz's inequality, we see that

$$[\xi(s_2) - \xi(s_1)]^2 \le \int_{s_1}^{s_2} \sin \pi s \, ds \int_{s_1}^{s_2} \frac{\xi'^2}{\sin \pi s} ds.$$

We integrate both sides of this inequality over the triangle $0 < s_1 < s_2 < 1$ and apply integration by parts on the right to find that

$$\int_0^1 \xi^2 \, ds - \left\{ \int_0^1 \xi \, ds \right\}^2 \le \frac{1}{\pi^2} \int_0^1 \xi'^2 \, ds.$$

This is the inequality (8.3) for the case $a = 0$, $b = 1$. As we pointed out at the beginning of the proof, a linear change of the variable of integration then yields (8.3) for any interval.

We remark that the inequality (8.3) is sharp in the sense that equality is attained when $\xi = \cos(\pi(s - a)/(b - a))$

To see how the Poincaré inequality is used, we consider the supremum problem (3.10) for the beam.

Break the interval $[0,1]$ into r subintervals of length $1/r$. By applying the Poincaré inequality on each of these intervals and adding the results, we find that the r conditions

$$\int_{(\nu-1)/r}^{\nu/r} \xi \, ds = 0 \qquad \text{for } \nu = 1, \cdots, r \tag{8.4}$$

imply that

$$\int_0^1 \xi^2 \, ds \le \frac{1}{\pi^2 r^2} \int_0^1 \xi'^2 \, ds.$$

Thus if the stiffness function $q(s)$ is bounded below by the positive constant q_{min}, we see that the linear constraints (8.3) imply that

$$\frac{\int_0^1 \xi^2 \, ds}{\int_0^1 q \xi'^2 \, ds} \le \frac{1}{\pi^2 q_{min} r^2}.$$

If r is sufficiently large, the right-hand side is less that the supremum in (3.10) so that the hypothesis of the Maximization Condition is satisfied.

We conclude that there is a maximizer ξ^* for the problem (3.10). The formal derivation in Section 3 then shows that this maximizer is the solution of the differential equation (3.11)

The same considerations justify the derivation of the differential equation (4.6) for the constrained problem (4.3).

We have actually proved that for the above problems making r sufficiently large makes the maximum of the Rayleigh quotient arbitrarily small. This property turns out to have other uses. We shall say that the quadratic form $Q[\xi]$ is **completely continuous** with respect to the positive definite quadratic form $R[\xi]$ if for every positive number ϵ there is a finite collection of constraints $\ell_1[\xi] = \cdots = \ell_r[\xi] = 0$ which implies that $|Q[\xi]| \leq \epsilon R[\xi]$.

The Poincaré inequality (8.3) for an interval is easily extended to a rectangle in two dimensions and to a parallelepiped in any number of dimensions. Roughly speaking, it serves to show that the supremum for a ratio of two quadratic integrals is attained if the integrand in the denominator involves higher derivatives of the function than the integrand in the numerator. The critical buckling load then appears as an eigenvalue of a partial differential operator.

The advantage of knowing that a maximizer for the problem (7.1) exists is two-fold. As we have seen, the existence statement often reduces the problem to one of finding the lowest eigenvalue of a differential operator. In one dimension there are efficient numerical algorithms to approximate this eigenvalue and to bound the error in this approximation.

We shall show in the next Section that the existence of a maximizer also permits us to bound the error made in the Rayleigh Ritz approximation. This leads to a lower bound for the critical buckling load.

It is interesting to know what happens when the hypothesis (8.1) is violated.

THEOREM. *There is no finite set of linear functionals for which (8.1) is satisfied if and only if there is an infinite sequence ξ_n with the properties that $R[\xi_n, \xi_m] = Q[\xi_n, \xi_m] = 0$ when $n \neq m$, $R[\xi_n] = 1$, and $\lim_{n\to\infty} Q[\xi_n] = \mu$, where μ is the supremum on the right of (8.1).*

Proof. Suppose first that there is no finite set of linear constraints $\ell_1[\xi] = \cdots = \ell_r[\xi] = 0$ which makes the supremum on the left of (8.1) less than the supremum μ on the right. By the definition of the supremum, there is an element ξ_1 of S for which $R[\xi_1] = 1$ and $Q[\xi_1] \geq \mu - 1$. Because the supremum is not reduced by imposing the linear constraints $R[\xi, \xi_1] = 0$ and $Q[\xi, \xi_1] = 0$, there is a ξ_2 which satisfies these constraints and for which $R[\xi_2] = 1$ and $Q[\xi_2] \geq \mu - \frac{1}{2}$. By proceeding in this fashion, we construct for every integer n a ξ_n which satisfies the linear constraints $Q[\xi_n, \xi_j] = R[\xi_n, \xi_j] = 0$ for $j < n$ such that $R[\xi_n] = 1$ and $Q[\xi_n] \geq \mu - \frac{1}{n}$. Then the sequence $\{\xi_n\}$ has the stated properties.

Assume, conversely, that one has a sequence with these properties. Choose any linear functionals $\ell_1, \cdots, \ell_r$. For any prescribed integer m there is a nontrivial linear combination ξ of the elements $\xi_m, \xi_{m+1}, \cdots, \xi_{m+r+1}$ which satisfies the constraints $\ell_1[\xi] = \cdots = \ell_r[\xi] = 0$. A simple computation which uses the properties of the sequence shows that $Q[\xi] \geq (\mu - (1/m))R[\xi]$. Therefore the supremum on the left of (8.1) is at least $\mu - (1/m)$. Since m is arbitrary, this shows that the condition (8.1) is not satisfied for any finite set of linear constraints.

9. An error bound for the Rayleigh-Ritz method.

We recall that the reciprocal of the critical buckling for a large class of problems is given as a supremum of a problem like (3.10). These problem have the form (7.1).

The Rayleigh-Ritz method always gives a lower bound for such a problem. Finding an upper bound for the supremum in (7.1) is equivalent to finding a bound for the error made in approximating the supremum by the Rayleigh-Ritz bound. We shall show how to do this under some special assumptions. Note that the reciprocal an upper bound for the supremum in (3.10) will give a safe load for the beam.

We first suppose that there is a maximizer ξ^* at which the supremum in (7.1) is attained. Then if η is any other element of S, the ratio in (7.1) with $\xi = \xi^* + \epsilon \eta$ has its maximum at $\epsilon = 0$. Therefore the derivative with respect to ϵ of the ratio vanishes at $\epsilon = 0$. If we let μ denote the supremum in (7.1), this fact translates into the generalized eigenvalue equation

$$(9.1) \qquad\qquad Q[\xi^*, \eta] = \mu R[\xi^*, \eta]$$

for all $\eta \in S$.

The Rayleigh-Ritz method consists of choosing p linearly independent elements of S, and computing

$$\mu^{[p]} := \max_{c_1, \cdots c_p} \frac{Q[\sum_1^p c_\nu \tau_\nu]}{R[\sum_1^p c_\nu \tau_\nu]}.$$

This $\mu^{[p]}$ is the largest eigenvalue of the matrix $Q[\tau_\mu, \tau_\nu]$ with respect to the matrix $R[\tau_\mu, \tau_\nu]$, and it is a lower bound for μ:

$$\mu \geq \mu^{[p]}.$$

We shall confine our attention to the case in which Q is positive definite. We first show that if one knows that one can approximate ξ^* well with a linear combination of the τ_ν, then $1/\mu^{[p]}$ is a good approximation to the reciprocal $1/\mu$ of the supremum.

THEOREM 9.1. *Suppose that Q and R are both positive definite, and that there is a ξ^* in S such that*

$$(9.2) \qquad\qquad \frac{Q[\xi^*]}{R[\xi^*]} = \mu := \sup_{\substack{\xi \in S \\ \xi \neq 0}} \frac{Q[\xi]}{R[\xi]}.$$

Let $\mu^{[p]}$ be the Rayleigh-Ritz lower bound for μ obtained with the trial functions $\tau_1, \cdots, \tau_p$. Then for any linear combination $\sum a_\nu \tau_\nu$ the inequalities

$$(9.3) \qquad \left(1 - \frac{R[\xi^* - \sum_{\nu=1}^p a_\nu \tau_\nu]}{R[\xi^*]}\right) \frac{1}{\mu^{[p]}} \leq \frac{1}{\mu} \leq \frac{1}{\mu^{[p]}}$$

are valid.

Proof. Let τ be that linear combination of the τ_ν which is nearest to the maximizer ξ^* in the sense of the norm $R^{1/2}$. That is, the coefficients of τ are chosen so that

$$R[\xi^* - \tau, \tau_\nu] = 0 \qquad \text{for } \nu = 1, \cdots, p.$$

Then

$$(9.4) \qquad\qquad R[\xi^*, \tau] = R[\tau],$$

and hence

$$(9.5) \qquad\qquad R[\xi^*] = R[\tau] + R[\xi^* - \tau].$$

We now see from (9.4) and (9.1) that

$$0 \leq Q[\tau - (R[\tau]/R[\xi^*])\xi^*] = Q[\tau] - 2\frac{R[\tau]}{R[\xi^*]}Q[\xi^*, \tau] + \frac{R[\tau]^2}{R[\xi^*]^2}Q[\xi^*] = Q[\tau] - \mu\frac{R[\tau]^2}{R[\xi^*]}$$

The definition of the Rayleigh-Ritz bound shows that $Q[\tau] \leq \mu^{[p]}R[\tau]$. Therefore the nonnegativity of the right-hand side shows that

$$\frac{1}{\mu} \geq \frac{R[\tau]}{R[\xi^*]}\frac{1}{\mu^{[p]}}.$$

We see from (9.5) that the ratio $R[\tau]/R[\xi^*]$ can be written as $1 - (R[\xi^* - \tau]/R[\xi^*])$. Thus we have established the first inequality in (9.3) for the special linear combination τ of the τ_ν . Since, by definition, $R[\xi^* - \sum a_\nu \tau_\nu] \geq R[\xi^* - \tau]$, this also establishes the inequality for all other linear combinations.

The second inequality in (9.3) is just the first monotonicity principle, so that the proof is complete.

Theorem 9.1 is of use only if one can show that a particular set of τ_ν has a linear combination τ which approximates the maximizer well in the sense that the relative error $R[\xi^* - \tau]/R[\xi^*]$ is small. If the dimension of the space S is greater than p, this property cannot hold for every element ξ^* of S, because one can choose ξ^* perpendicular to the τ_ν to make the best approximation $\tau = 0$. We must therefore make use of the special property (9.1) of the maximizer.

We shall not discuss a general theory, but we shall show how $R[\xi^* - \sum a_\nu \tau_\nu]$ can be made small in a particular problem.

We recall that the Maximization Condition of Section 8 can be used to show that the beam problem (3.10) has a maximizer ξ^*. The derivation in Section 3 shows that this maximizer is a solution of the boundary value problem (3.11).

We assume that the stiffness function q is continuously differentiable, and that there are positive constants q_{min} and q_{max} such that

$$q_{min} \leq q(s) \leq q_{max}.$$

We also assume the bound

$$|q'(s)| \leq c$$

for some constant c. Then we see from (3.11) and the triangle inequality that (9.6)

$$\left\{ \int_0^1 |\xi^{*\prime\prime}|^2 ds \right\}^{1/2} \leq \frac{c}{q_{min}^{3/2}} R[\xi^*]^{1/2} + \frac{P^*}{q_{min}} Q[\xi^*]^{1/2} = \left\{ \frac{c}{q_{min}^{3/2}} + \frac{\sqrt{P^*}}{q_{min}} \right\} R[\xi^*]^{1/2}.$$

Thus the maximizer ξ^* has the special property that one can bound the integral of the square of its second derivative in terms of $R[\xi^*]$. We now recall the "rooftop functions", which were introduced in Section 7. The linear combinations of these functions are continuous and piecewise linear, with jumps in the derivative only at the mesh points $s = j/p$. In particular, the second derivative of such a linear combination is zero in each subinterval between these points. Hence, the function $\xi^* - \sum a_\nu \tau_\nu$ has the same second derivatives as ξ^* in each of the subintervals.

We now choose the particular linear combination which has the property that

$$\int_{(j-1)/p}^{j/p} [\xi^* - \sum a_\nu \tau_\nu]' ds = 0$$

for $j = 1, \cdots, p$. Because ξ^* and the τ_ν vanish at $s = 0$, this condition simply states that the linear combination has the same values as ξ^* at the mesh points.

By the Poincaré inequality of Section 8, we see that

$$\int_{(j-1)/p}^{j/p} [\xi^* - \sum a_\nu \tau_\nu]'^2 ds \leq \frac{1}{\pi^2 p^2} \int_{(j-1)/p}^{j/p} [\xi^* - \sum a_\nu \tau_\nu]''^2 ds$$

Because the τ_ν are piecewise linear, their second derivatives vanish on each subinterval. We add these inequalities to obtain the bound

$$R[\xi^* - \sum a_\nu \tau_\nu] \leq \frac{q_{max}}{\pi^2 p^2} \int_0^1 |\xi^{*\prime\prime}|^2 ds.$$

We substitute (9.6) in this inequality to find that

$$R[\xi^* - \sum_{\nu=1}^{p} a_\nu \tau_\nu] \leq \frac{q_{max}}{\pi^2 p^2} \left\{ \frac{c}{q_{min}^{3/2}} + \frac{\sqrt{P^*}}{q_{min}} \right\}^2 R[\xi^*].$$

We substitute this inequality in the lower bound (9.3) for $(1/\mu) = P^*$ to find the inequality

$$P^* \geq \left(1 - \frac{q_{max}}{\pi^2 p^2} \left\{ \frac{c}{q_{min}^{3/2}} + \frac{\sqrt{P^*}}{q_{min}} \right\}^2 \right) \frac{1}{\mu^{[p]}}.$$

P^* appears on both sides of this inequality. However, a simple use of the quadratic formula yields an explicit lower bound of the form $\{1 + O(p^{-2})\}/\mu^{[p]}$ for the critical buckling load P^*.

We have thus shown that if the Rayleigh-Ritz method is applied with the piecewise linear functions $\tau_1, \cdots, \tau_p$, then while the Rayleigh-Ritz bound $1/\mu^{(p)}$ gives an unsafe load, one can construct a safe load which is very near to $1/\mu^{(p)}$ when p is sufficiently large. This idea can be applied to many other problems in one or more dimensions with a variety of choices of the Rayleigh-Ritz functions.

STABILITY OF DISCRETE SYSTEMS

H.R. Milner

Monash University, Melbourne, Australia

ABSTRACT

This paper describes the general features of structural stability problems and pays particular attention to eigen-analysis. Both non-linear eigen solutions are presented and linear eigen solutions which arise out of minimum total potential energy formulations such as arise with the finite element method. By examination of the properties of Rayleigh Quotients various summation and bounding theorems can be derived from these linear eigen solutions and are presented towards the end of the paper.

1 PRELIMINARIES

1.1 Strain Definitions

Stability theory is a sub-set of the theory of elasticity, sharing with it a common set of postulates and definitions. Central to these theories are relationships between stress and deformation (strain) discovered by subjecting real materials to stress and observing their response; the best known of these relationships is Hooke's law.

In the process of establishing stress-strain relationships, the definition of *engineering strain* emerged intuitively as the *change in length per unit original length* (see Appendix A, eqn A.1). It provides (a) a linear relationship between strain and filament elongation (a form of geometric linearity) and (b) many materials obey Hooke's law at low to moderate strain levels. Unfortunately, once strains are expressed in terms of displacements, the geometric linearity persists only in one dimension unless small displacement geometry is applicable. For larger strains alternative deformation measures, eg, Green's strain (eqn A.2) can become at least equally appealing. For the strain levels met in stability problems, *Green's strain* is practically identical to *engineering strain* and often leads to simpler algebra.

Although stability theory deals with the effect of secondary displacements on structural behaviour, it does not normally deal with materials and components subjected to large strains. Hence it is satisfactory to utilise, interchangeably, any of the strain definitions described in the literature. The matter is addressed more completely in Appendix A.

1.2 Function Maxima and Minima

In stability theory it will often be necessary to determine whether a function, $f(x)$, is maximum or minimum. This determination is commonly made by examining a Taylor's series expansion of $f(x)$ in the region of x, see eqns 1.1a and 1.1b.

$$f(x+\Delta x) = f(x) + \Delta x f'(x) + \frac{\Delta x^2}{2!} f''(x) + \frac{\Delta x^3}{3!} f'''(x) + \ldots\ldots \frac{\Delta x^n}{n!} f^n(x) + \ldots. \qquad 1.1a$$

or

$$\Delta f = f(x+\Delta x) - f(x) = \Delta x f'(x) + \frac{\Delta x^2}{2!} f''(x) + \frac{\Delta x^3}{3!} f'''(x) + \ldots\ldots \frac{\Delta x^n}{n!} f^n(x) + \ldots. \qquad 1.1b$$

By taking Δx sufficiently small, the first term on the right side of eqn 1.1b becomes dominant and Δf has the same sign as $\Delta x f'(x)$ which changes sign with Δx. A necessary condition for a stationary point, where the function is momentarily unchanged for small variations of either sign of the independent variable x, is $f'(x) = 0$. Thus, at a stationary point, expression 1.1b reduces to

$$\Delta f = \frac{\Delta x^2}{2!} f''(x) + \frac{\Delta x^3}{3!} f'''(x) + \ldots\ldots + \frac{\Delta x^n}{n!} f^n(x) + \ldots \qquad 1.1c$$

It follows that, if $f''(x)$ is negative, $f(x)$ is maximum and, if $f''(x)$ is positive, $f(x)$ is minimum; it is always possible to reduce Δx so that the first term in expression 1.1c is dominant. If $f''(x)$ vanishes with $f'(x)$, which occurs in some buckling problems, then higher order terms in the Taylor's series expansion must be examined. Suppose that the highest non-vanishing derivative in the expansion occurs for a derivative of order m; expression 1.1c then takes the form

$$\Delta f = \frac{\Delta x^m}{m!} f^m(x) + \dots \qquad \text{1.1d}$$

where $m > 1$.

If m is odd, the expression 1.1d changes sign with Δx and $f(x)$ is neither maximum or minimum at x; it is a point of inflection. If m is even, the function is maximum if $f^m(x)$ is negative and minimum if $f^m(x)$ is positive.

Where this principle is applied to functions of total potential energy, Π, it takes the form

$$\Delta\Pi = \Pi(D + \Delta D) - \Pi(D) = \Delta D\Pi'(D) + \frac{\Delta D^2}{2!}\Pi''(D) + \frac{\Delta D^3}{3!}\Pi'''(D) +$$
$$\dots + \frac{\Delta D^n}{n!}\Pi^n(D) + \dots \qquad \text{1.2a}$$

where $D =$ displacement. For equilibrium, $\Pi'(D) = 0$ and thus

$$\Delta\Pi = \frac{\Delta D^2}{2!}\Pi''(D) + \frac{\Delta D^3}{3!}\Pi'''(D) + \dots + \frac{\Delta D^n}{n!}\Pi^n(D) + \dots \qquad \text{1.2b}$$

If the total potential energy is a function of several generalised displacements, the positive definiteness of the partial derivatives needs to be investigated, ie, the D in eqn 1.2 is interpreted as a vector.

2   THE NATURE OF STRUCTURAL INSTABILITY

As put succintly by Croll and Walker (1972), the development of new materials and structural forms has resulted, over the last 200 years, in a trend towards light, thin-walled metal structures. Our ability to construct such structures has been brought about mainly by the development of high strength metals and other materials but this has not always been accompanied by corresponding increases in modulus of elasticity. As a consequence, instability phenomena, which are strongly influenced by material modulus of elasticity, have become more prominent and a strong interest has developed in the topic.

Our understanding of structural instability has developed principally along two parallel paths. General stability theories have become known through the writings of Poincare (1885),

Koiter (1945), Croll and Walker (1972), Thompson and Hunt (1973). Simultaneously, extensive literature on the bifurcation buckling of specific structural forms, notably columns, beams, frameworks, plates and shells has become available to engineers through the writings of Bleich (1952) and Timoshenko and Gere (1961) in particular. More recently, a variety of authors have taken advantage of the advances in computer technology and the development of finite difference and finite element methods to study the behaviour of specific structures over the pre and post post-buckling ranges. Extensive data is now available for a wide range of structural forms and is expressed in design codes and standard texts as approximate formulae, tables and charts. Because such knowledge is so widely available in a simplified form, eg, the design of I beam webs under a variety of in plane stresses, the majority of engineers require relatively little knowledge of general stability theories to undertake routine design.

The material covered herein is directed at that part which leads to the formulation of eigen-problems. Such problems occur in what Murray (1984) refers to as a first order stability analysis described by Weinberger (1994) as linear stability analysis. As will be seen later, the evaluation of stresses which cause the buckling effects are based on a zero order or small deflection analysis. As such, eigen analysis can only determine a bifurcation buckling load; it cannot analyse structural behaviour in the post-buckling range; second or higher order analyses are required for that purpose.

To introduce general stability theory and to place eigen-analysis in context, simple structures, called link structures by Thompson and Hunt (1973), are usually studied. The treatment of this general theory herein is necessarily brief; for more complete descriptions the reader is referred to the readily available texts of Croll and Walker (1972) and Thompson and Hunt (1973). Both books are very readable introductions to general stability theories.

2.1   Spring Supported Rigid Bar

Simple, one degree of freedom systems have been used by many writers to demonstrate the general features of stability problems. One of the simplest problems of this type arises with a rigid bar having infinite bending stiffness and a rotational spring at its lower end; see Fig 2.1. The load, P, is vertical, remains so during lateral movement and travels laterally with the bar tip. A zero order stability analysis ignores completely any instability effects, assuming, in effect, that the bar always remains vertical.

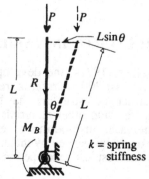

Fig 2.1 Rigid bar - linear rotational spring system.

2.1.1 First Order Analysis - Bar Supported by a Linear Spring

Statical Equilibrium Approach

A first order stability analysis of this structure requires that the analyst have the capacity to perceive equilibrium configurations of the form illustrated by the dotted line in Fig 2.1. Small displacement geometry is used, eg, ($\theta \approx \sin \theta$), which provides only limited information about the stability of the structure. If $k = constant$ = rotational spring stiffness, then, for small angles of rotation, θ, the bar will remain in equilibrium if

$$PL\theta - k\theta = 0 \qquad\qquad 2.1$$

Two solutions are possible

- $R = P = arbitrary\ value, \theta = 0$,
- $R = P = k/L, \ \theta = arbitrary\ value$

The $\theta = arbitrary\ value$ solution is possible only if $P = k/L$, otherwise the $P = arbitrary\ value, \theta = 0$ solution applies.

Force Residual Approach

It is also possible to introduce a moment, M_B, at the base which represents the difference between the spring reaction and the over-turning moment caused by the axial load (the out-of-balance force). For fixed θ there is only one value of P which corresponds to an equilibrium value, ie, when $M_B = 0$. The value of M_B value is related to P and θ by

$$M_B = PL\theta - k\theta = (PL - k)\theta \qquad\qquad 2.2$$

If, for positive θ, M_B is positive, then it is assisting the base spring in resisting the over-turning moment caused by the load P; if, under such circumstances, the base moment was removed, the angle θ would increase and the structure would move away from an equilibrium position. Such a movement away from an equilibrium position is characteristic of instability behaviour. The reverse behaviour characterises stability since the base moment would be assisting the load, P, in overcoming the spring reaction and its removal would result in the structure returning to its equilibrium position. Applying these principles to the rigid bar of Fig 2.1 and considering eqn 2.2, it follows that, if $PL < k$, then a positive change in θ requires a negative change in M_B and the structure is stable. By a similar argument, it follows that, if $PL > k$, then a positive change in θ requires a positive change in M_B and the structure is unstable.

When $PL = k$ the system is in neutral equilibrium. If disturbed from its equilibrium position under these conditions, the increased base moment caused by load, P, exactly counter-balances the increased base moment reaction generated by the spring. Eqn 2.2 predicts that the system behaves like a piece of putty and will stay in whatever position it is placed by an external disturbance.

Alternatively, reference can be made to the derivative of M_B with respect to θ. Positive $dM_B/d\theta$ indicates instability, negative $dM_B/d\theta$ stability and $dM_B/d\theta = 0$ indifferent (putty-like) equilibrium. To use this approach, eqn 2.2 is differentiated with respect to θ which leads to

$$\frac{dM_B}{d\theta} = PL - k \qquad\qquad 2.3$$

and the previous conclusions redrawn.

Total Potential Energy Approach

Finally, it is possible to adopt a total potential energy perspective which classifies the stability of equilibrium on the basis of its total potential energy status as follows:

status of total potential energy	stability classification
minimum	stable
maximum	unstable
indifferent	neutral

In the problem of Fig 2.1, the total potential energy is given by

$$\Pi(\theta) = \tfrac{1}{2}k\theta^2 - \tfrac{1}{2}PL\theta^2 \qquad\qquad 2.4$$

A Taylor's series expansion of the energy expression can now be examined, viz,

$$\Delta\Pi = \Delta\theta\Pi'(\theta) + \frac{\Delta\theta^2}{2}\Pi''(\theta) + \ldots = \Delta\theta(k\theta - PL\theta) + \frac{\Delta\theta^2}{2}(k - PL) \qquad 2.5$$

A state of equilibrium exists if $\Pi'(\theta) = 0$ and this leads to the equation of equilibrium given as eqn 2.1, viz,

$$k\theta - PL\theta = 0 \qquad\qquad 2.6$$

This has a trivial solution $\theta = 0$, $P = arbitrary\ value$. The stability of the equilibrium along this load path can be considered for particular values of P and $\theta = 0$. By substituting $\Pi'(\theta) = 0$ and $\theta = 0$ into eqn 2.5, it reduces to

$$\Delta\Pi = \frac{\Delta\theta^2}{2}(k - PL) \qquad\qquad 2.7$$

Eqn 2.7 shows that the equilibrium is stable for $PL < k$, unstable for $PL > k$ and indifferent for $PL = k$. All higher derivatives are zero.

2.1.2 Higher Order Analysis - Linear Spring

Statical Equilibrium Approach

The statements given in Section 2.1.1 provide a limited description of the system behaviour given that small geometry approximations break down at finite displacement levels. If finite displacements are considered, the equation of equilibrium takes the modified form

$$PL\sin\theta - k\theta = 0 \qquad\qquad 2.8a$$

$$\text{or} \qquad P = \frac{k\theta}{L\sin\theta} \qquad\qquad 2.8b$$

Eqn 2.8a has a trivial solution $\theta = 0$, $P = k/L$ (note that $\theta/\sin\theta \to 1$ as $\theta \to 0$) but there exists non-trivial solutions for non-zero θ values. The solution of eqn 2.8b is plotted in non-dimensionalised form in Fig 2.2.

Total Potential Energy Approach

The total potential energy, Π, is given by

$$\Pi(\theta) = \tfrac{1}{2}k\theta^2 - PL(1 - \cos\theta) \qquad\qquad 2.9$$

The Taylor's series expansion of the total potential energy function is given by

$$\Delta\Pi = \Delta\theta(k\theta - PL\sin\theta) + \Delta\theta^2(k - PL\cos\theta)$$
$$+ \Delta\theta^3(PL\cos\theta) + \Delta\theta^4(PL\cos\theta) + \ldots\ldots\ldots \qquad\qquad 2.10a$$

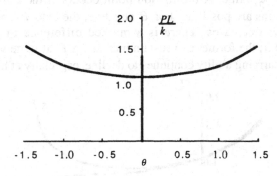

Fig 2.2 Bar - linear spring system behaviour for large deformations.

For the structure to be in equilibrium, $\Pi'(\theta) = k\theta - PL\sin\theta = 0$. Substituting $PL = k\theta/\sin\theta$ into eqn 2.10b, it reduces to

$$\Delta\Pi = \Delta\theta^2 k(1 - \theta/\tan\theta) + \Delta\theta^3(k\theta) + \Delta\theta^4(k\theta/\tan\theta) + \ldots\ldots \qquad\qquad 2.10b$$

Except at the point $\theta = 0$, the term $\Delta\theta^2 k(1 - \theta/\tan\theta)$ and therefore $\Delta\Pi$, is positive in the range $-\pi < \theta < \pi$. At the point $\theta = 0$ the terms $\Delta\theta^2 k(1 - \theta/\tan\theta)$ and $\Delta\theta^3(k\theta)$ also vanish but $\Delta\theta^4(k\theta/\tan\theta)$ is positive. This establishes the minimum character of the stationary point.

The path $(0 \leq PL/k \leq 1, \ \theta = 0)$ is known as the fundamental load path. At the point $(PL/k = 1, \ \theta = 0)$ the structure has a choice of load paths said to represent a *bifurcation point*. The load at which it occurs is known as the *critical load*. The magnitude of the critical load can be predicted by either a first or higher order stability analysis.

2.1.3 Higher Order Analysis - Non-linear Spring

<u>Statical Equilibrium Approach</u>

If the case of a non-linear spring, the behaviour in the post-buckling range is strongly influenced by the spring stiffness characteristics. If, for example, the spring stiffness is given by $k(\theta + k_1\theta^2)$, then equilibrium configurations can be expressed in the form

$$PL\sin\theta = k(\theta + k_1\theta^2) \qquad\qquad 2.11a$$
$$\frac{PL}{k} = \frac{\theta + k_1\theta^2}{\sin\theta} \qquad\qquad 2.11b$$

Numerical results based on eqn 2.11b are shown in Fig 2.3 for $k_1 = -1/3$ and $k_1 = -1/3.5$. As occurred in the linear spring case, $\theta = 0$ represents an equilibrium position for any value of the load, P, for any k_1 value. A bifurcation point occurs at the critical load $PL/k = 1$ where non-trivial solutions are possible. For θ negative, the load P continues to increase and, for θ positive, P decreases. There is a marked difference in the behaviour for $k_1 = -1/3$ and $k_1 = -1/3.5$; the former can sustain increasing P at large values of θ but, in the latter case, the load carrying ability continues to decline, especially at higher values of θ.

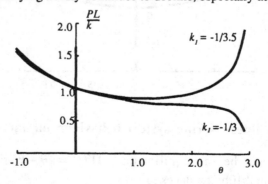

Fig 2.3 Bar - non-linear spring system behaviour for large deformations.

Force Residual Approach

The structure can also be analysed in the manner of Section 2.1.1, eqn 2.2 with an out-of-balance base moment included as

$$M_B = PL\sin\theta - k(\theta + k_1\theta^2) \qquad 2.12$$

To complete this analysis the derivative is taken, namely,

$$\frac{dM_B}{d\theta} = PL\cos\theta - k(1 + 2k_1\theta)$$

$$= PL(1 - \theta^2/2 + ...) - k(1 + 2k_1\theta) \qquad 2.13$$

$$= (PL - k) - PL\,\theta^2/2 - 2kk_1\theta$$

The stability of the system can be examined as before. The remarks made earlier about the stability of the trivial solution can be repeated here. The non-trivial solution becomes possible at load $PL/k = 1$; its stability is examined by investigating the behaviour of eqn 2.13. Since the system must be in equilibrium, eqn 2.11a can be substituted into eqn 2.13 leading to

$$\frac{dM_B}{d\theta} = \frac{k(\theta + k_1\theta^2)}{\tan\theta} - k(1 + 2k_1\theta) \qquad 2.14$$

An algebraic examination of eqn 2.14 is possible but a numerical evaluation is simpler and more instructive, see Fig 2.4. The equilibrium is stable where $dM_B/d\theta$ is negative and unstable where it is positive.

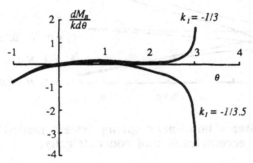

Fig 2.4 Rate of change of extraneous base moment required to maintain equilibrium.

2.1.4 Large Displacement Analysis - Non-linear Spring - Eccentric Load

A further case of interest occurs if the load is applied with eccentricity $e = NL$ where $N =$ ratio of eccentricity to length of the rigid bar. The behaviour of the system is described by the expression

$$\frac{PL}{k} = \frac{\theta + k_1\theta^2}{\sin\theta + N\cos\theta}$$

2.15

For the purpose of illustration, it is assumed that either $N = 0$ or $N = 0.05$, which produces the forms of behaviour illustrated in Fig 2.5. Consider the case of structure softening, $k_1 = -1$, in the post-buckling range. With the eccentrically loaded member, $N = 0.05$, a maximum load given by $PL/k = 0.644$ at $\theta = 0.211$ is reached at which point $dP/d\theta = 0$. At large angles, the load-rotation curve for the eccentrically loaded bar-spring system asymptotes that of the concentrically loaded bar-spring system. The reason for this behaviour is that the eccentricity progressively makes proportionally less and less of a contribution to the over-turning moment.

The general characteristics displayed by the three types of behaviour illustrated in Fig 2.5 are displayed to some degree by more recognizable structural components:

(a)   the $k_1 = 1$ behaviour is similar in form to the response of a rectangular shaped steel plate panel loaded in shear,

(b)   the $k_1 = 0$ behaviour is similar in form to the response of a column made from steel with an infinite yield point,

(c)   the $k_1 = -1$ behaviour is similar in form to the response of a column made from steel with a finite yield point.

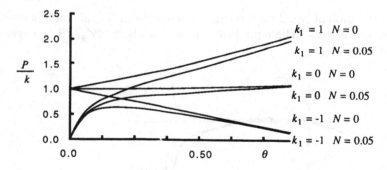

Fig 2.5 Bar - non-linear spring system loaded
eccentrically and concentrically.

2.1.5 Large Displacement Analysis - Non-linear Spring - Initial Disturbance

Consider a bar not vertical at the commencement of loading. Let the bar of Fig 2.1 have an initial clockwise rotation, θ_0, which causes no reactive force in the spring and base spring stiffness $= k(\theta + k_1\theta^2)$; it is only the additional rotation caused by load application which causes base spring reaction. The equation of equilibrium is given by

$$PL\sin(\theta + \theta_0) = k(\theta + k_1\theta^2)$$ 2.16

After rearrangement it can be shown that

$$\frac{PL}{k} = \frac{1}{\sin\theta_0}\left(\frac{\theta + k_1\theta^2}{\sin\theta + \tan\theta_0\cos\theta}\right) = g\left(\frac{\theta + k_1\theta^2}{\sin\theta + N_1\cos\theta}\right)$$ 2.17

where $g = constant = 1/\sin\theta_0$.

Eqn 2.17 is formally similar to eqn 2.15 which shows that lack of initial verticality has a similar effect to eccentric load application.

2.1.6 A Two Degree of Freedom System

Elsewhere in this publication Weinberger (1994) has analysed a two degree of freedom problem identical to the one illustrated in Fig 2.6. The analysis is repeated briefly herein for the sake of completeness.

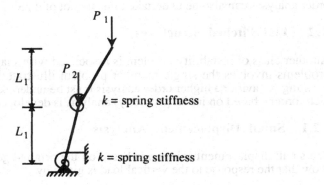

Fig 2.6: Two degree of freedom rigid bar structure.

The zero order, total potential energy statement has the form

$$\Pi = k\theta_1^2 - k\theta_1\theta_2 + \tfrac{1}{2}k\theta_2^2 - \tfrac{1}{2}P_1L_1\theta_1^2 - \tfrac{1}{2}P_2L_1\theta_1^2 - \tfrac{1}{2}P_2L_2\theta_2^2$$ 2.18

The eigen-problem is obtained by setting $\Delta\Pi = 0$ or the partial derivatives of Π with respect to θ_1 and θ_2 to zero. Thus

$$\frac{\partial \Pi}{\partial \theta_1} = 2k\theta_1 - k\theta_2 - P_1 L_1 \theta_1 - P_2 L_1 \theta_1 = 0 \qquad \qquad 2.19a$$

$$\frac{\partial \Pi}{\partial \theta_2} = -k\theta_1 + k\theta_2 - P_2 L_2 \theta_2 = 0 \qquad \qquad 2.19b$$

For a fixed load pattern and loads proportional to a load factor, λ, eqns 2.19 have the form

$$\left[\begin{bmatrix} 2k & -k \\ -k & k \end{bmatrix} - \lambda \begin{bmatrix} P_1 L_1 + P_2 L_1 & 0 \\ 0 & P_2 L_2 \end{bmatrix} \right] \begin{bmatrix} \theta_1 \\ \theta_2 \end{bmatrix} = \begin{bmatrix} 0 \\ 0 \end{bmatrix} \qquad \qquad 2.20$$

Letting $k = 1$, $L_1, L_2 = 1$

$$\left[\begin{bmatrix} 2 & -1 \\ -1 & 1 \end{bmatrix} - \lambda \begin{bmatrix} P_1 + P_2 & 0 \\ 0 & P_2 \end{bmatrix} \right] \begin{bmatrix} \theta_1 \\ \theta_2 \end{bmatrix} = \begin{bmatrix} 0 \\ 0 \end{bmatrix} \qquad \qquad 2.21$$

Once the ratio $P_1 : P_2$ is prescribed and base values provided, the eigen-pairs can be extracted. The reader is referred to the discussion by Weinberger (1994) for further details. Higher order analyses can also be undertaken but are not of direct interest herein.

2.2   Flat Pitched Structures

Another class of instability problem is associated with snap buckling. The simplest of such problems involves the single member problem illustrated in Fig 2.7. To predict the snap buckling behaviour a higher order analysis must be undertaken although a small displacement (zero order - based on initial geometry) analysis is developed first.

2.2.1   Small Displacement Analysis

The small displacement solution is based on the original geometry; it is a trivial exercise to show that the response to the vertical load is given by

$$P = F \sin\theta = \frac{EA}{\sqrt{l^2 + h^2}} (\sin^2 \theta) d \qquad \qquad 2.22$$

This solution predicts that the bar can support an indefinitely large force at a sufficiently large displacement, d.

2.2.2 Large Displacement Analysis (Based on Green's Strain)

A somewhat diferent behaviour is predicted if large displacement effects are included in the analysis. Suppose that the bar has the initial length, l_0, given by

$$l_0^2 = l^2 + h^2 \qquad \qquad 2.23$$

and a length after displacement, l_1, given by

$$l_1^2 = l^2 + (h-d)^2 \qquad\qquad 2.24$$

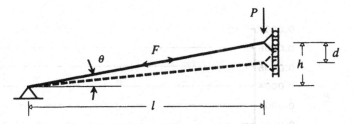

Fig 2.7 Flat pitched structure.

Assuming that Green's strain is used, its magnitude is given by,

$$\varepsilon_G = -\frac{l_1^2 - l_0^2}{2l_0^2} = -\frac{d^2 - 2dh}{2(l^2 + h^2)} \qquad\qquad 2.25$$

where
h = height of the bar above horizontal when unloaded
d = displacement from the unloaded position, positive downwards.

The minus sign is introduced so that compression strain is positive.

Because the bar is in equilibrium, the axial force, F, load, P, and displacement, d, are related by

$$P = F \sin\theta$$

$$= F \frac{h-d}{\sqrt{l^2 + (h-d)^2}} \qquad\qquad 2.26$$

$$= -EA\left[\frac{d^2 - 2dh}{2(l^2 + h^2)}\right]\frac{h-d}{\sqrt{l^2 + (h-d)^2}}$$

Numerical results for $A=1$, $l=10$, $h=1$, are plotted in Fig 2.8; it represents the load, P, which can be resisted for specific values of displacement, d.

Consider circumstances in which the load, P, is increased from zero. When the ratio, P/E, reaches a value of approximately 0.00019023 the vertical component of the bar compression force can no longer increase; the only stable configuration involves the bar snapping across into tension. Once the bar is in tension, its resistance to downwards force, P, increases so long as material strength permits. If the force reverses in direction the bar can be forced (snapped) back into its original configuration by a force $P/E \geq 0.00019023$.

It is possible to examine the stability of this system by the same methods used to examine the stability of the bar with the base spring. In this case an extraneous vertical force is place under the the load point and the rate of change of this force with displacement is noted. Such analysis will show that 0A and CB are stable load paths but that AC is an unstable load path.

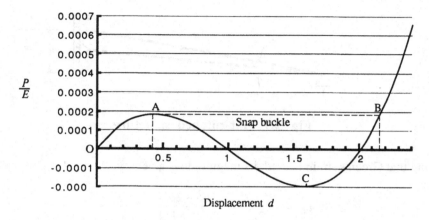

$\frac{P}{E}$

Displacement d

Fig 2.8 Behaviour of the structure illustrated in Fig 2.7   The load path forms a mirror image for P negative.

2.3 Summary of Buckling Types

The terms "buckling", "load limit instability" and "bifurcation instability" have all been used to date to describe a number of phenomena. Herein, the term "buckling" will be used as a general term to describe both load-limit instability and bifurcation instability.

2.3.1 Bifurcation Instability

The type of buckling illustrated in Section 2.1 leads to a bifurcation of equilibrium. If the stability analysis is first order, then it also leads to an eigen-problem. Initially, the structure responds to the applied load by deforming in what is termed the *fundamental displacement mode*. On the load-displacement diagram of Fig 2.9 the fundamental displacement mode is associated with the *fundamental load path*. In the structures described in Section 2.1 the *fundamental displacement mode* involves the rigid bar remaining vertical.

At the *critical load* or *bifurcation point*, an alternative deformation pattern becomes possible which is termed the *post-buckling displacement mode*. On the load-displacement diagram of Fig 2.9 the post-buckling displacement mode is associated with the *post-buckling load path*. At the *bifurcation point*, the structure can resist the load in either of the two displacement modes. (The word bifurcation is defined in the Webster dictionary as "a division into two branches".)

Following the writings of many workers including Koiter (1945, 1974), Thompson and Hunt (1973), Chen and Lui (1986), the type of bifurcation is also classified by the shape of the post-buckling load path as stable and symmetric, see Fig 2.10(a), unstable and symmetric, see Fig 2.10(b), asymmetric, see Fig 2.10(c), or neutral, Fig 2.10(d). True neutral bifurcation is very unusual in practice although apparent neutral equilibrium can arise in some analyses due to the fact that some higher order non-linear effects are over-looked, eg, with the first order Euler column solution discussed in Section 3.1. As drawn, Fig 2.9 represents a neutral equilibrium form in which, as discussed earlier, the behaviour is putty-like, ie, the structure will remain in displaced form in which it is placed by an external agency so long as the critical load continues to act.

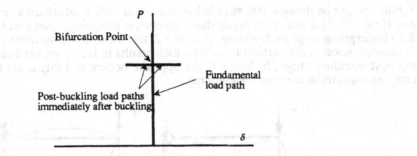

Fig 2.9 Load-displacement diagram illustrating bifurcation of equilibrium.

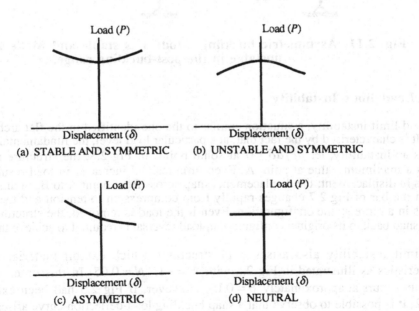

Fig 2.10   Forms of bifurcation as classified by their post-buckling behaviour.

With symmetric bifurcation, either stable or unstable post-buckling behaviour occurs depending on whether or not the post-buckling load paths result in increasing or decreasing total potential energy. With asymmetric behaviour, the load carrying capacity increases along one post-buckling load path and decreases along the other.

While it might be thought that such behaviour arises only in idealised structures, Thompson and Hunt (1973) actually point out that asymmetric behaviour arises with the frame of Fig 2.10 undergoing in-plane buckling. In Mode 2, the shear in the horizontal member opposes the compression in the vertical member which results in increased load carrying capacity in the post-buckling range. In Mode 1, the opposite occurs and this leads to a reduced load carrying capacity in the post-buckling range.

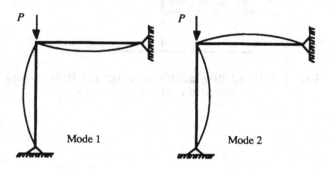

**Fig 2.11 Asymmetric buckling. Mode 1 is stable and Mode 2
unstable in the post-buckling range.**

2.3.2 Load limit Instability

With load limit instability, a situation similar to the one identified in the flat arch situation arises. It is characterised by the fact that, at a particular load level, the fundamental load path exhibits an instability, ie, $dP/d\delta = 0$ at some point. In Fig 2.8, the structure resistance reaches a maximum value at point A. Even infinitesimal increases in load result in large changes in displacement; the displacements snap across from point A to B. In this case, the force in the bar of Fig 2.7 changes rapidly from compression to tension and the structure deforms in a more stable configuration. Even if the load is removed, the structure will not readily snap back to its original configuration; load reversal is required to achieve that result.

Load limit instability also arises with structures which exhibit material softening characteristics as illustrated in Fig 2.5 with $k_1 = -1$, $N = 0.05$. In this case, load limit instability occurs at approximately $\theta = 0.15$. However, if Fig 2.5 had been extended to $\theta > \pi/2$, it is possible to observe that a snap buckling load-deflection curve arises with the rigid rod supporting the load, P, by acting in tension.

3   BIFURCATION - CLASSICAL FIRST ORDER STABILITY ANALYSES OF COMMON STRUCTURAL ELEMENTS

3.1   Euler Column Buckling

In the classical Euler column problem, a perfectly straight bar is subjected to pure axial compression; no primary bending moments are applied. This response of the column to axial compression is described by the differential equation

$$EA(dw/dz) + P = 0 \qquad\qquad 3.1\text{a}$$

and boundary condition $w = 0$ at $z = 0$. It has the general solution

$$w = z(P/EA) \qquad\qquad 3.1\text{b}$$

where
$E =$ elastic modulus,
$A =$ cross-section area,
$z =$ longitudinal axis,
$w =$ axial displacement in direction of the z axis,
$P =$ axial compression.

Eqns 3.1(a) and 3.1(b) predict that the column will remain perfectly straight and continue to carry unlimited axial load. Such a solution represents a load path which always remains stable no matter how the analysis is viewed.

The column can also be investigated to determine if alternative equilibrium configurations exist involving bending actions described in accordance with the differential equation,

$$EI(d^4v/d^4z) + P(d^2v/d^2z) = 0 \qquad\qquad 3.1(\text{c})$$

where
$v =$ bending displacement,
$I =$ second moment of area for the direction of buckling.

Equation 3.1(c) is applicable to any beam-column subject to axial load applied either concentrically or eccentrically. It does not cover situations in which transverse load is applied. If the load is applied eccentrically with zero end moment, then $v = e_1, e_2$, $d^2v/dz^2 = 0$ at $z = 0, L$ and eqn 3.1 has the solution

$$v = A + Bz + C\sin(kz) + D\cos(kz) \qquad\qquad 3.1(\text{d})$$

where $k^2 = P/EI$. Because attention will be focussed herein on concentrically loaded columns, it will be assumed that $v = 0$ at $z = 0, L$. After introduction of these constraints, eqn 3.1(d) becomes

$$v = A\left(1 - \frac{z}{L} - \cos kz + \frac{z}{L}\cos kL\right) + C\left(\sin kz - \frac{z}{L}\sin kL\right) \qquad\qquad 3.1(\text{e})$$

The constants A, C in eqn 3.1(e) can be chosen to arrive at the deflected shape of any column subject to concentric axial load only. A number of first order buckling solutions may be obtained. Example 3.1 introduces typical cases.

Example 3.1 Solution of the Differential Equation for Various Boundary Conditions

Eqn 3.1(e) has the form

$$v = A\left(1 - \frac{z}{L} - \cos kz + \frac{z}{L}\cos kL\right) + C\left(\sin kz - \frac{z}{L}\sin kL\right)$$

By differentiation

$$\frac{dv}{dz} = A\left(-\frac{1}{L} + k\sin kz + \frac{1}{L}\cos kL\right) + C\left(k\cos kz - \frac{1}{L}\sin kL\right)$$

$$\frac{d^2v}{dz^2} = A\left(k^2\cos kz\right) - C\left(k^2\sin kz\right)$$

Pinned - Pinned Column

The boundary conditions are $d^2v/dz^2 = 0$ at $z = 0, L$ which leads to the expressions

$$Ak^2 \qquad\qquad\qquad = 0 \qquad\qquad\qquad\qquad \text{A}$$

$$Ak^2\cos(kL) \quad -Ck^2\sin(kL) = 0 \qquad\qquad\qquad \text{B}$$

Equations A and B have a trivial solution corresponding to the fundamental load path where $v = 0$, $P = arbitrary\ value$. This load path is stable for $kL < \pi$ and unstable for $kL > \pi$. The non-trivial solution is possible for $k^4\sin kL = 0$, or, $kL = n\pi$, where n is a positive integer. This leads to the well-known Euler solution for a pin ended column, $P_{cr} = \pi^2EI/(nL)^2$ which has a minimum value $P_{cr} = \pi^2EI/L^2$. A problem of this type is a non-linear eigen-problem since the critical loads are extracted as non-linear functions of a characteristic polynomial. The load paths are illustrated in Fig 3.1.

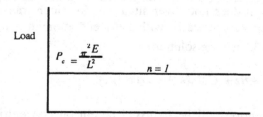

Fig 3.1 Illustration of bifurcation at critical loads of a perfectly straight, pin-ended column.

The buckled shape can be obtained by noting, from eqn A, that $A = 0$ and $\sin kL = 0$ is a necessary condition for a solution to exist. Hence the buckled shape is given by $v = C \sin kz$. It is not possible to assign a specific numerical value to C.

Pinned - Fixed and Fixed - Fixed Solutions
It is left as an exercise to verify that the eigen-values for the above cases are obtained from the following expressions

Pinned - Fixed Case $\tan kL = kL$
The first two roots are given by $kL = (4.4934 \quad 7.7253)$

Fixed - Fixed Case $2 - kL \sin kL - 2 \cos kL = 0$
The first two roots are given by $kL = (6.2832 \quad 8.9868)$

3.2   Lateral-torsional Buckling of Beams

The lateral-torsional buckling case discussed most frequently involves the application of pure uniform bending moment about the major axis of a doubly-symmetric I beam. Such members are susceptible to a form of buckling involving twist and minor axis bending. This behaviour is described by the differential equations,

$$EI_x \frac{d^2 v}{dz^2} - M = 0 \qquad \qquad 3.2a$$

$$EI_y \frac{d^2 u}{dz^2} - \phi M = 0 \qquad \qquad 3.2b$$

$$GJ \frac{d\phi}{dz} - EI_w \frac{d^3 \phi}{dz^3} + \frac{du}{dz} M = 0 \qquad \qquad 3.2c$$

where
I_y = minor axis second moment of area
J = St Venant's torsion constant
I_w = warping constant
u = minor axis bending deflections
ϕ = angle of twist
M = bending mement applied about the major axis
E, G = elastic constants

Eqns 3.2b and 3.2c, which deal with the buckling action, can be combined into a single equation involving only the angle of twist, ϕ. This has the form

$$EI_w \frac{d^4 \phi}{dz^4} - GJ \frac{d^2 \phi}{dz^2} - \frac{M_0^2}{EI_y} \phi = 0 \qquad \qquad 3.3$$

After making the substitutions

$$\alpha = GJ/2EI_w, \quad \beta = M_0^2/E^2 I_y I_w, \quad m = \sqrt{-\alpha + \sqrt{\alpha^2 + \beta}}, \quad n = \sqrt{\alpha + \sqrt{\alpha^2 + \beta}},$$

the general solution of equation eqn 3.3 has the form,

$$\phi = A\sin(mz) + B\cos(mz) + Ce^{nz} + De^{-nz} \qquad 3.4$$

An eigen-problem is formed by using the boundary conditions. If the conditions $\phi, d\phi/dz = 0$ at $z = 0, L$ are taken, then it can be shown that $B = 0$, $C = -D$ from the conditions at $z = 0$ and that $\phi = A\sin(mz) - 2D\sinh(nz)$. Using the boundary conditions at $z = L$, the expressions,

$$A\sin(mL) \quad - 2D\sinh(nL) = 0$$
$$Am^2 \sin(mL) + 2Dn^2 \sinh(nL) = 0 \qquad 3.5$$

are obtained. Eqns 3.4 have a non-trivial solution only if,

$$\det \begin{vmatrix} \sin(mL) & -2\sinh(nL) \\ m^2 \sin(mL) & 2n^2 \sinh(nL) \end{vmatrix} = 0 \qquad 3.6$$

Expansion of the determinant leads to the conclusion that $\sin(mL) = 0$ or $mL = \pi$. After substitution for m, n, α, β it follows that

$$M_{cr} = \frac{\pi}{L} \sqrt{EI_y GJ \left(1 + \frac{EI_w}{GJ} \frac{\pi^2}{L^2}\right)} \qquad 3.7$$

4   INSTABILITY - THE ENERGY PERSPECTIVE

4.1   Theorem of Virtual Displacements for a Column derived directly from the Beam Equation of Equilibrium

The theorem of virtual displacements is developed in the Appendix B for a beam member subject only to transverse loads, end moment and end shear. Such actions cause internal bending and shear reactions only - no direct force reactions. If the longitudinal axis is labelled z the theorem statement takes the form

$$\int_0^L M_i \Delta\phi dz = M_1 \Delta\theta_1 + M_2 \Delta\theta_2 + V_1 \Delta v_1 + V_2 \Delta v_2 + \int_0^L q\Delta v dz \qquad 4.1$$

Eqn 4.1 is applicable in the case of a beam satisfying the equations of equilibrium

$$\frac{d^2M_i}{dz^2} - q = 0$$

$$V_i = -\frac{dM_i}{dz}$$

4.2

$$q = -\frac{dV_i}{dz}$$

The sign convention for internal actions (bending moments and shears) and boundary actions (transverse force and end moments) are reproduced in Fig 4.1 for convenience.

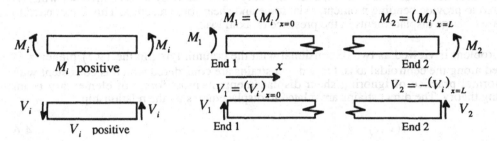

(a) Positive internal bending moments and shears

(b) Positive boundary tractions

Fig 4.1   Sign convention for internal actions and end actions.

If the member is also subject to axial load, P, taken as positive, if compressive, and the displacements small, the equations of equilibrium become, Timoshenko (1961),

$$\frac{d^2M_i}{dz^2} + P\frac{d^2v}{dz^2} - q = 0$$

$$V_i = -\frac{dM_i}{dz} + P\frac{dv}{dz}$$

4.3

$$q = -\frac{dV_i}{dz}$$

The theorem of virtual work is using integration by parts and takes the form

$$\int_0^L (M_i + Pv)\Delta\phi \, dz = \left[(M_i + Pv)\Delta\phi\right]_0^L - \int_0^L \frac{d}{dz}(M_i + Pv)\left(\frac{d\Delta v}{dz}\right)dz$$

$$= \left[(M_i + Pv)\Delta\phi\right]_0^L + \int_0^L V_i\left(\frac{d\Delta v}{dz}\right)dz$$

$$= \left[(M_i + Pv)\Delta\phi + V_i\Delta v\right]_0^L + \int_0^L q\Delta v \, dz$$

4.4

At the ends, it is assumed that the axial load is applied concentrically so that the expression reduces to

$$\int_0^L (M_i + Pv)\Delta\phi dz = \left[M_i\Delta\theta + V_i\Delta v \right]_0^L + \int_0^L q\Delta v dz \qquad 4.5$$

which is the virtual displacement statement for Euler buckling.

4.2   Theorem of Minimum Potential Energy for a Column

The theorem of minimum potential energy is also developed in the Appendix B for the case where transverse loads, end moment and shear only (no axial load) act on a member which is required to provide bending moment, axial force and shear force reaction. This is extended to the case of large displacements in the presence of axial load.

The problem is viewed as two dimensional with the column lying in the (y, z) plane; z is aligned along the centroidal axis. Only direct strains are considered which is consistent with our normal practice of ignoring shear displacements in most forms of elementary beam bending theory. The direct strains are related to displacements by the relationship

$$v(y', z) = v \qquad 4.6$$

$$w(y', z) = w - y'\frac{dv}{dz} \qquad 4.7$$

where

$v(y', z) = v$ displacement at height y' above the centroidal axis

$w(y', z) = w$ displacement at height y' above the centroidal axis

$v =$ buckling displacement of the centroidal axis

$w = w$ displacement of the centroidal axis

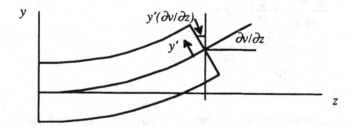

Fig 4.2   **Contribution to w displacements from beam slope.**

Using eqn A.18

$$\varepsilon_z = \frac{dw}{dz} + \frac{1}{2}\left(\frac{dv}{dz}\right)^2 + \frac{1}{2}\left(\frac{dw}{dz}\right)^2 \qquad 4.8$$

Substituting from eqns 4.6 and 4.7 into eqn 4.8 leads to

$$\varepsilon_{zz}(y',z) = \frac{dw}{dz} - y'\left(\frac{d^2v}{dz^2}\right) + \frac{1}{2}\left(\frac{dv}{dz}\right)^2$$
$$+ \frac{1}{2}\left(\frac{dw}{dz}\right)^2 - y'\left(\frac{dw}{dz}\right)\left(\frac{d^2v}{dz^2}\right) + \frac{1}{2}y'^2\left(\frac{d^2v}{dz^2}\right)^2 \qquad 4.9$$

The last three terms of eqn 4.9 are negligible by comparison with the other terms so that the simplified expression 4.10 is adopted.

$$\varepsilon_{zz} = \frac{dw}{dz} - y'\frac{d^2v}{dz^2} + \frac{1}{2}\left(\frac{dv}{dz}\right)^2 \qquad 4.10$$

The strain energy is given by

$$U = \frac{1}{2}\int_V E\varepsilon_{zz}^2 dV \qquad 4.11$$

Substituting eqn 4.10 into eqn 4.11 and writing $dV = Adz$ leads to

$$U = \frac{1}{2}\int_L\int_A\left[\left(\frac{dw}{dz}\right)^2 + y'^2\left(\frac{d^2v}{dz^2}\right)^2 + \frac{1}{4}\left(\frac{dv}{dz}\right)^4 - 2y'\left(\frac{dw}{dz}\right)\left(\frac{d^2v}{dz^2}\right)\right.$$
$$\left. - y'\left(\frac{d^2v}{dz^2}\right)\left(\frac{dv}{dz}\right)^2 + \left(\frac{dw}{dz}\right)\left(\frac{dv}{dz}\right)^2\right]EdAdz \qquad 4.12$$

After integration across the depth and using the relationships $\int_A dA = A$, $\int_A y'dA = 0$, $\int_A y'^2 dA = I$, the strain energy is finally given by

$$U = \frac{1}{2}\int_L\left[A\left(\frac{dw}{dz}\right)^2 + I\left(\frac{d^2v}{dz^2}\right)^2 + A\left(\frac{dw}{dz}\right)\left(\frac{dv}{dz}\right)^2 + \frac{A}{4}\left(\frac{dv}{dz}\right)^4\right]Edz \qquad 4.13$$

Compression loads are taken to be positive, which means that critical loads will be positive; the critical loads would be negative if the sign convention for axial load was reversed. After substituting $P_i = EA(dw/dz)$, omitting $(A/4)(dv/dz)^4$ as being a negligible and separating the terms containing v and w displacements the strain energy is given as $U = U_v + U_w$, where

$$U_w = \frac{1}{2}\int_L EA\left(\frac{dw}{dz}\right)^2 dz \qquad 4.14$$

$$U_v = \frac{1}{2}\int_L\left[EI\left(\frac{d^2v}{dz^2}\right)^2 - P_i\left(\frac{dv}{dz}\right)^2\right]dz \qquad 4.15$$

The total potential energy associated with the bending deformations only is given by or

$$\Pi_v = 0.5\int_0^L \left[EI\phi^2 - P_i(dv/dz)^2\right]dz - M_1\theta_1 - M_2\theta_2 - V_1v_1 - V_2v_2 - \int_0^L qvdz \quad 4.16(a)$$

$$\Pi_v = 0.5\int_0^L \left[EI\phi^2 - P_i(dv/dz)^2\right]dz - \left[M_i\theta + V_iv_i\right]_0^L - \int_0^L qvdz \qquad 4.16(b)$$

Considering the first variation in potential energy

$$\Delta\Pi_v = \int_0^L \left[EI\phi\delta\phi - P_i(dv/dz)(d(\delta v)/dz)\right]dz - \left[M_i\delta\theta + V_i\delta v_i\right]_0^L - \int_0^L q\delta vdz \quad 4.17$$

Integrating the second term in brackets by parts, assuming that the axial load is applied concentrically at the ends and substituting $M_i = EI\phi$ leads to

$$\Delta\Pi_v = \int_0^L \left[M_i\Delta\phi + P_iv\Delta\phi\right]dz - \left[M_i\Delta\theta + V_i\Delta v_i\right]_0^L - \int_0^L q\Delta vdz \qquad 4.18$$

In view of the theorem of virtual work

$$\Delta\Pi_v = \int_0^L \left[M_i\Delta\phi + P_iv\Delta\phi\right]dz - \left[M_i\Delta\theta + V_i\Delta v_i\right]_0^L - \int_0^L q\Delta vdz = 0 \qquad 4.19$$

A second variation of the total potential energy can be taken leading to

$$\Delta(\Delta\Pi_v) = \Delta^2\Pi_v = \int_0^L \left[EI(\Delta\phi)^2 + P_i\Delta v\Delta\phi\right]dz \qquad 4.20$$

If $\Delta^2\Pi_v$ is positive, the total potential energy is minimum and the column stable but, if negative, the total potential energy is maximum and the column unstable. Expression 4.20 can be used to detect load limit instability and neutral or indifferent equilibrium.

4.2 Summary of Work Product and Energy Theorems

The theorem of virtual displacements is less restrictive than the theorem of minimum total potential energy. The former only implies that an equilibrium force field has been postulated but the latter also incorporates a constitutive equation and requires that the displacements are statically admissible. Both theorems can be used interchangeably and the theorem of virtual work can be converted to a statement which is equivalent to the theorem minimum total potential energy by insisting that the displacements are actual rather than virtual and introducing a constitutive equation. Some writers of finite element texts only ever use the theorem of virtual work. The additional constraints involving a constitutive law and meeting kinematical admissibility conditions are introduced as part of deriving element characteristics. A problem often exists with finite elements meeting kinematical admissibility requirements at the inter-element boundaries.

5   FINITE ELEMENT FORMULATION OF BUCKLING PROBLEMS

The finite element formulations of buckling problems which follow are based on the work of Gallagher (1975); the principle of minimum strain energy is employed in their derivation.

5.1   In-plane Buckling

5.1.1   Element Stiffness and Geometric Matrices

In the finite element method as it applies to column buckling, the bending displacements within an element are expressed in the form $v = a + bz + cz^2 + dz^3$. The terms a, b, c, d are replaced by the nodal variables v_1, θ_1, v_2, θ_2. The substitutions $v_1 = a$, $\theta_1 = b$, $v_2 = a + bL + cL^2 + dL^3$, $\theta_2 = b + 2cL + 3dL^2$ and $\xi = z/L$ are used to obtain, after some algebra ,

$$v = \left[(1 - 3\xi^2 + 2\xi^3) \quad (\xi - 2\xi^2 + \xi^3)L \quad (3\xi^2 - 2\xi^3) \quad (-\xi^2 + \xi^3)L \right] \begin{Bmatrix} v_1 \\ \theta_1 \\ v_2 \\ \theta_2 \end{Bmatrix} \quad 5.1$$

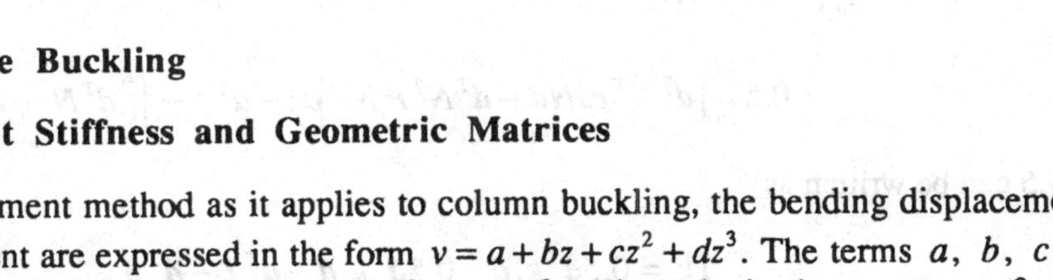

**Fig 5.1: Element parameters and displacement geometry
for column buckling.**

Eqn 5.1 may be written in matrix notation as

$$v = \begin{bmatrix} N_1 & N_2 & N_3 & N_4 \end{bmatrix} \begin{Bmatrix} v_1 \\ \theta_1 \\ v_2 \\ \theta_2 \end{Bmatrix} = [N]\{d\} \quad 5.2$$

Bold type is used to indicate a matrix quantity. By differentiation, it follows that curvature, ϕ, and slope, dv/dz are given by

$$\phi = \ddot{N}d \quad 5.3$$

$$dv/dz = \dot{N}d \quad 5.4$$

The dots indicate differentation with respect to the longitudinal (z) axis. Writing the element end actions as $f^T = \langle V_1 \quad M_1 \quad V_2 \quad M_2 \rangle$ and substituting eqns 5.3 and 5.4 into the theorem of total minimum potential energy leads to

$$\Pi_v = 0.5 \int_0^L \left[EI\phi^2 - P_i (dv/dz)^2 \right] dz - M_1\theta_1 - M_2\theta_2 - V_1v_1 - V_2v_2 - \int_0^L qvdz$$

$$= 0.5 \int_0^L \left[d^T \ddot{N}^T EI\ddot{N}d - d^T \dot{N}^T P_i \dot{N}d \right] dz - d^T f - \int_0^L d^T N^T qdz \qquad 5.5$$

Eqn 5.5 can be written as

$$\Pi_v = 0.5 d^T k_f d - 0.5 d^T k_g d - d^T f - d^T p \qquad 5.6$$

where

$$k_f = \int_0^L EI\ddot{N}^T \ddot{N} dz \qquad k_g = \int_0^L P_i \dot{N}^T \dot{N} dz \qquad p = \int_0^L N^T qdz \qquad 5.7$$

Using eqn 5.1 the stiffness matrix, k_f, is given by

$$k_f = \begin{bmatrix} \dfrac{12EI}{L^3} & & symmetric & \\ \dfrac{6EI}{L^2} & \dfrac{4EI}{L} & & \\ \dfrac{-12EI}{L^3} & \dfrac{-6EI}{L^2} & \dfrac{12EI}{L^3} & \\ \dfrac{6EI}{L^2} & \dfrac{2EI}{L} & \dfrac{-6EI}{L^2} & \dfrac{4EI}{L} \end{bmatrix} \qquad 5.8$$

and the geometric matrix, k_g, by,

$$k_g = \frac{P_i}{30L} \begin{bmatrix} 36 & & symmetric & \\ -3L & 4L^2 & & \\ -36 & 3L & 36 & \\ -3L & -L^2 & 3L & 4L^2 \end{bmatrix} \qquad 5.9$$

5.1.2 Global Stiffness and Geometric Matrices and Solution Techniques

Eqns 5.6 - 5.9 are used to develop a finite model for a complete structure by noting that the total potential energy of the complete structure as an assemblage of elements is simply the sum of the total potential energies of the elements. As it applies to the complete structure, eqn 5.7 takes the form of eqn 5.10. In eqn 5.10 the upper case letters indicate that the quantities refer to the complete structure. In the process of summing the term $d^T f$ cancels except on the boundaries leaving finally

$$\Pi_{v(total)} = 0.5D^T K_f D - 0.5D^T K_g D - D^T P \qquad 5.10$$

Taking the first variation leads to

$$\Delta\Pi_{v(total)} = 0 = \Delta D^T K_f D + \Delta D^T K_g D - \Delta D^T P \qquad 5.11$$

The terms ΔD^T are arbitrary, so that eqn 5.11 can be satisfied only if

$$\left(K_f - K_g\right)D - P = 0 \qquad 5.12$$

Eqn 5.12 can be used to analyse a framework in which beam-column type actions are operative. This includes both the destiffening effect of compressive axial forces and the stiffening effect of axial tension. The 4×4 matrices k_f and k_g and their global counter-parts can be expanded to 6×6 matrices by including axial load stiffness coefficients (EA/L).

5.1.3 Eigen-analysis

If the second variation of eqn 5.11 is taken and set to zero to detect a bifurcation instability this leads to

$$\Delta^2\Pi_{v(total)} = \Delta D^T K_f \Delta D - \Delta D^T K_g \Delta D = 0 \qquad 5.13a$$

or, since the terms ΔD^T are arbitrary,

$$\left(K_f - K_g\right)\Delta D = 0 \qquad 5.13b$$

Eqn 5.13b is tantamount to an eigen-analysis statement with the matrix K_g containing terms proportional to the axial load, P_i, acting on individual members. In an eigen-analysis, it is assumed that these member axial loads (stress resultants) remain constant under proportional loading. A "load factor", λ, is sought by which the stress resultants can be multiplied to obtain non-trivial solutions of eqn 5.13b. Let the load factor be λ so that the eqn 5.13b becomes

$$\left(K_f - \lambda K_g\right)\Delta D = 0 \qquad 5.14$$

Discussion

The "load factor" computed by eqn 5.14 is determined relative to some base level of loading. Although the term "load factor" is widely used it is probably misleading. The term "stress factor" is probably better since it is the internal stress resultants which give rise to the instability effects.

In any statically determinate truss or frame, the stress resultants (axial loads, shears and bending moments) are independent of element stiffness properties, ie, "load factor" and "stress factor" lead to the same result.

On the other hand, in statically indeterminate trusses or frameworks, the member axial loads P_i are functions of element stiffness properties. A number of solution techniques can be adopted to arrive at the base stress distribution which is to be factored by λ in any subsequent eigen-analysis. The most common basis is obtained from a zero order analysis as follows.

i)   Compute the stresses (axial loads) using eqn 5.12 with all $P_i = 0$ and thus all k_g null.

ii)   Compute the k_g matrices with the P_i values determined from (i).

iii)   Compute λ_{cr} and the eigen-modes from eqn 5.14.

It is also possible to undertake an iterative procedure in which successive k_g matrices are determined by iteration on eqn 5.12; the procedure is continued until the displacements stabilise or diverge. Displacement stabilisation will always occur if the frame is loaded below its smallest critical load factor and diverge if above this level. Such a process can be continued with the applied loads progressively increased until the displacements diverge during the iterations so that an eigen-analysis is never performed. It is also possible to perform an eigen-analysis at the completion of any iterative step which is then regarded as the basic stress distribution.

Example 5.1 Euler Column Problem - Finite Element Analysis
**   - Structure Statically Determinate**

Consider a two element solution of the Euler column buckling problem for which the displacement vector is $d = \langle v_1 \quad \theta_1 \quad v_2 \quad \theta_2 \quad v_3 \quad \theta_3 \rangle^T$. The axial load is known from statics. If only symmetric buckling modes (those having an odd number of half waves and forming a mirror image in the two halves of the column) are extracted, the displacement vector is reduced to $d = \langle v_1 \quad \theta_1 \quad v_2 \quad \theta_2 \rangle^T$. The stiffness matrices for the left element only are used in the analysis. Such modes are subject to the further geometric boundary conditions $v_1, \theta_2 = 0$. Writing $P_i = P_{cr}$, $L = 0.5l$, the characteristic equation becomes

$$\left(\frac{8EI}{l} - \frac{P_{cr}l}{15} \right)\left(\frac{96EI}{l^3} - \frac{12P_{cr}}{5l} \right) - \left(\frac{24EI}{l^2} - \frac{P_{cr}}{10} \right)^2 = 0$$

There are two positive solutions $P_{cr} = 9.944\left(\dfrac{EI}{l^2} \right)$ and $P_{cr} = 128.723\left(\dfrac{EI}{l^2} \right)$ representing approximations to an Euler column buckling in one and three half waves respectively; the precise answers are $P_{cr} = 9.870\left(\dfrac{EI}{l^2} \right)$ and $P_{cr} = 88.826\left(\dfrac{EI}{l^2} \right)$.

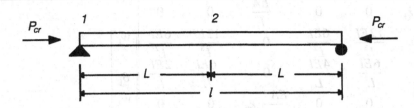

Fig 5.2   Euler problem solved by finite element techniques.

Example 5.2 Euler Column Problem - Finite Element Analysis
- Structure Statically Indeterminate

This example is taken from a paper by Tarnai (1981) and reference will be made to it later. It involves an Euler strut with springs having stiffness k_1 and k_2 at the two ends and with vertical load applied at the upper spring-column junction.

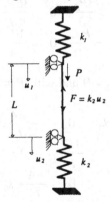

Fig 5.3 Structure where the critical load is decreased by increasing the stiffness of some components.

Supplementing the stiffness and geometric matrices given in eqns 5.8 and 5.9 by including terms which relate to axial deformation, the structure response can be expressed in the form of eqn 5.12 as

$$
\begin{bmatrix}
\dfrac{EA}{L}+k_1 & 0 & 0 & -\dfrac{EA}{L} & 0 & 0 \\[2mm]
0 & \dfrac{12EI}{L^3} & \dfrac{6EI}{L^2} & 0 & -\dfrac{12EI}{L^3} & \dfrac{6EI}{L^2} \\[2mm]
0 & \dfrac{6EI}{L^2} & \dfrac{4EI}{L} & 0 & -\dfrac{6EI}{L^2} & \dfrac{2EI}{L} \\[2mm]
-\dfrac{EA}{L} & 0 & 0 & \dfrac{EA}{L}+k_2 & 0 & 0 \\[2mm]
0 & -\dfrac{12EI}{L^3} & -\dfrac{6EI}{L^2} & 0 & \dfrac{12EI}{L^3} & -\dfrac{6EI}{L^2} \\[2mm]
0 & \dfrac{6EI}{L^2} & \dfrac{2EI}{L} & 0 & -\dfrac{6EI}{L^2} & \dfrac{4EI}{L}
\end{bmatrix}
\begin{bmatrix} w_1 \\ v_1 \\ \theta_1 \\ w_2 \\ v_2 \\ \theta_2 \end{bmatrix}
$$

$$
-\begin{bmatrix}
0 & 0 & 0 & 0 & 0 & 0 \\[2mm]
0 & \dfrac{36P_i}{30L} & -\dfrac{3P_i}{30} & 0 & -\dfrac{36P_i}{30L} & -\dfrac{3P_i}{30} \\[2mm]
0 & -\dfrac{3P_i}{30} & \dfrac{4P_iL}{30} & 0 & \dfrac{3P_i}{30} & -\dfrac{P_iL}{30} \\[2mm]
0 & 0 & 0 & 0 & 0 & 0 \\[2mm]
0 & -\dfrac{36P_i}{30L} & \dfrac{3P_i}{30} & 0 & \dfrac{36P_i}{30L} & \dfrac{3P_i}{30} \\[2mm]
0 & -\dfrac{3P_i}{30} & -\dfrac{P_iL}{30} & 0 & \dfrac{3P_i}{30} & \dfrac{4P_iL}{30}
\end{bmatrix}
\begin{bmatrix} w_1 \\ v_1 \\ \theta_1 \\ w_2 \\ v_2 \\ \theta_2 \end{bmatrix}
=
\begin{bmatrix} P \\ 0 \\ 0 \\ 0 \\ 0 \\ 0 \end{bmatrix}
$$

Imposing the boundary conditions $v_1, v_2 = 0$ and with $P_i = 0$ initially, it follows that

$$
\begin{bmatrix}
\dfrac{EA}{L}+k_1 & 0 & -\dfrac{EA}{L} & 0 \\[2mm]
0 & \dfrac{4EI}{L} & 0 & \dfrac{2EI}{L} \\[2mm]
-\dfrac{EA}{L} & 0 & \dfrac{EA}{L}+k_2 & 0 \\[2mm]
0 & \dfrac{2EI}{L} & 0 & \dfrac{4EI}{L}
\end{bmatrix}
\begin{bmatrix} w_1 \\ \theta_1 \\ w_2 \\ \theta_2 \end{bmatrix}
=
\begin{bmatrix} P \\ 0 \\ 0 \\ 0 \end{bmatrix}
$$

Assuming that $EA/L \gg k_1, k_2$, it is found that $w = w_1 = w_2 = P/(k_1 + k_2)$, $\theta_1 = \theta_2 = 0$. For the column, $P_i = k_2 w = P/(1 + k_1/k_2)$ which provides the stress resultant upon which a subsequent eigen-analysis is based. As the column is the only component which buckles the eigen-analysis can be undertaken with displacements $\Delta\theta_1, \Delta\theta_2$ according to eqn 5.14. Thus

$$
\left[
\begin{bmatrix}
\dfrac{4EI}{L} & \dfrac{2EI}{L} \\[2mm]
\dfrac{2EI}{L} & \dfrac{4EI}{L}
\end{bmatrix}
-
\begin{bmatrix}
\dfrac{4P_iL}{30} & -\dfrac{P_iL}{30} \\[2mm]
-\dfrac{P_iL}{30} & \dfrac{4P_iL}{30}
\end{bmatrix}
\right]
\begin{bmatrix} \Delta\theta_1 \\ \Delta\theta_2 \end{bmatrix}
=
\begin{bmatrix} 0 \\ 0 \end{bmatrix}
$$

which has, as its minimum critical load, $P_i = \dfrac{12EI}{L^2} = \dfrac{P_{cr}}{\left(1 + k_1/k_2\right)}$. This can be rearranged as

$P_{cr} = \dfrac{12EI}{L^2}\left(1 + k_1/k_2\right)$. A precise solution gives $P_{cr} = \dfrac{\pi^2 EI}{L^2}\left(1 + k_1/k_2\right)$.

5.2   Computerising the Solution - Jacobi Diagonalisation

It is possible to use standard eigen-solvers to extract the eigen-pairs and, subsequently, to use digital computers to produce numerical results. A simple solver, suitable for relatively small problems, is Jacobi diagonalisation. With this algorithm, the matrices are transformed, iteratively, to diagonal form.

Let A and B be two, positive-definite or negative-definite, symmetric matrices and suppose that the eigen-pairs $\left(\lambda_i, e_i\right)$ for the problem $(A - \lambda B)d = 0$ are required. The diagonalisation involves transformations of the form $A_{n+1} = T_n^T A_n T_n$ $B_{n+1} = T_n^T B_n T_n$ where T_n is a unit matrix modified by placing entries α and γ in the positions ij and ji where α and γ are given by

$$\alpha = M/P \qquad \gamma = -N/P \qquad\qquad 5.15$$
$$M = a_{ii}b_{ij} - b_{ii}a_{ij} \qquad N = a_{jj}b_{ij} - b_{jj}a_{ij}$$
$$Q = a_{ii}b_{jj} - b_{ii}a_{jj} \qquad P = \tfrac{1}{2}Q + sgn(Q)\sqrt{\left(\tfrac{1}{2}Q\right)^2 + MN} \qquad 5.16$$

A single transformation sets the terms a_{ij} and b_{ij} to zero but, once set to zero, they can become non-zero in subsequent transformations; i and j are chosen to correspond with the largest off-diagonal coefficient of A. Any banding of the matrices, such as arises with normal finite element analysis, is lost in the transformations.

After several such transformations, the matrices A and B tend to diagonalise; the process is terminated when the off-diagonal terms are reduced below a specified lower limit. Subsequently, the eigen-values can be calculated using $\lambda_i = a_{ii}/b_{ii}$ and the eigen-vectors using $E = T_1 T_2 T_3 T_4 \ldots T_n$. The eigen-vectors appear as columns in E. The matrix T_n is given specifically by

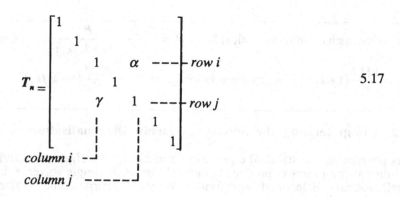

$$5.17$$

Depending on the sign conventions chosen, there are buckling problems in which the eigen-values are negative, eg, in the Euler problem, if positive load is tensile. This presents no difficulty as the Jacobi method still converges for negative definite systems. The only difficulty arises if matrix K_g is indefinite, ie, if some eigen-values are positive and some are negative. The problem can ordinarily be converted to a positive definite form by using a shifting procedure. If the system $(A - \lambda B)d = 0$ has eigen-values, λ, then an equivalent problem

$$[(A + \beta B) - \alpha B]d = 0 \qquad\qquad 5.18$$

where $\lambda = \alpha - \beta$, which may be solved to extract positive eigen-values only.

The positive constant, β, alters the matrix, A, in such a way that the eigen-values have an amount β added to them; β needs to be at least as large as the largest negative eigen-value. The problem becomes ill-conditioned if β is made too large; the method breaks down under such circumstances. In the following demonstration examples involving negative eigen values, β has been selected manually by trial and error.

In closing this section it is pointed out that Jacobi diagonalisation is not an efficient method for extracting the eigen-pairs of large systems. It suffers the following disadvantages:

(i)   it destroys the banding of matrices common in structural systems,
(ii)   it extracts all eigen-pairs whether or not these are required.

On the other hand, the computer code is easy to write. Alternative algorithms are available which are described elsewhere, eg, Bathe and Wilson (1972) and Mohr and Milner (1989). Algorithms such as vector iteration combined with deflation allow prescribed eigen-values to be extracted, eg, the lowest three.

5.3   Euler Column Problem Solved by Jacobi Iteration

Using the Jacobi technique, the Euler column problem of Fig 5.1 can be computed using the stiffness matrices, eqns 5.15 and 5.16; no negative eigen-values are involved in Euler buckling. The solution below was extracted with the rigid body modes included, ie, with the freedoms $u_1, \theta_1, u_2, \theta_2, u_3, \theta_3$ all present.

The results obtained are as follows.

Mode	1	2	3	4	5	6
Eigen-values	0.000	0.000	9.944	48.000	128.723	240.000
Exact			9.870	39.478	88.826	157.914
Eigen-vector						
u_1	0.5774	0.2060	0.0092	0.1028	0.0381	0.0009
θ_1	0.0000	-0.2619	-0.6421	0.5681	-0.7056	0.5774
u_2	0.5774	0.4680	0.4188	0.1028	-0.0356	0.0009
θ_2	0.0000	-0.2619	0.0000	0.5681	0.0000	0.5774
u_3	0.5774	0.7299	0.0092	0.1028	0.0381	0.0009
θ_3	0.0000	-0.2619	0.6421	0.5681	0.7056	0.5774

Modes 1 and 2 are rigid body modes representing, respectively, a pure lateral translation and a combined lateral translation and rotation. These appear in the solution because rigid body modes have not been eliminated by appropriate fixing of the supports. The value is finally determined by examining the nature of the eigen-vector; a rigid body movement without strain can only produce a zero eigen-value. Modes 3 through 6 represent approximations to the buckling load of an Euler column in 1, 2, 3, 4 half-waves. The eigen-values become progressively less accurate as the number of half-waves increases but this is unimportant in practice as it is the lower critical loads (eigen-values) which are of practical interest.

A feature of this solution is that all of the eigen-values are greater than or equal to zero which means that buckling is possible only if a compressive load is applied. Systems of this type are said to be positive-semi-definite.

5.5 Lateral-Torsional Beam Buckling by Finite Element Analysis

5.5.1 Expression for Total Potential Energy

It is shown by Bleich (1952) that the total potential energy for a prismatic, singly symmetric beam, subject to bending moment, caused by both applied end moment and transverse load, and axial load is given by

$$\Pi = \tfrac{1}{2}\int_0^L [EI_y \ddot{u}^2 + GJ\dot{\phi}^2 + EI_w \ddot{\phi}^2 - \frac{P_i I_p}{A}\dot{\phi}^2 - P_i \dot{u}^2$$

$$+ 2M\ddot{u}\phi - \bar{a}q\phi^2 + \left(F_{y1}z - F_{y2}(L-z) + M_{x1} - M_{x2}\right)\phi\ddot{u}\,]dz \qquad 5.19$$

$$- \sum_{i=1}^{2}(M_{zi}\phi + M_{zzi}\dot{\phi} + F_{xi}u_i + M_{yi}\dot{u}_i)$$

J = St Venant torsion constant

I_p = polar moment of inertia

I_w = warping constant
A = cross-section area
I_y = minor axis second moment of area
$\bar{a}$ = height above the shear centre of transverse load
$M = M_q + P_i y_0$
M_q = bending caused by q as if the beam was simply supported

The end actions are illustrated in Fig 5.4 and the transverse load in Fig 5.5.

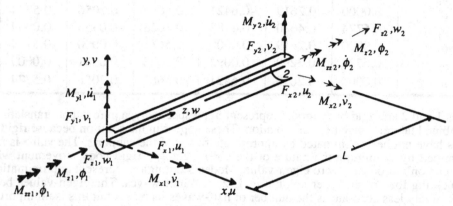

**Fig 5.4: Notation used for end actions of lateral-torsional beam
buckling problem.**

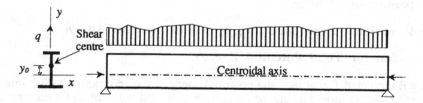

**Fig 5.5 Transverse load and shear centre for
lateral-torsional buckling.**

5.5.2 Finite Element Expressions

Gallagher (1976) has developed finite element expressions for lateral-torsional buckling for
the case of zero transverse load. The buckling displacements are minor axis deflection, u,
and twist, ϕ. There are two boundary conditions at each end in each of these variables, the
variable itself and its first derivative. Denoting the shape functions by N, the displacements
are given by

$$u = [N]\langle u_1 \quad \dot{u}_1 \quad u_2 \quad \dot{u}_2 \rangle^T \qquad \phi = [N]\langle \phi_1 \quad \dot{\phi}_1 \quad \phi_2 \quad \dot{\phi}_2 \rangle^T \qquad 5.20$$

Substitution of these into the energy functional leads to the stiffness expression

$$\{f\} = \left[[k_f] + [k_g]\right]\{d\} \qquad 5.21$$

where

$$\{f\} = \left[F_{z1} \quad M_{y1} \quad M_{z1} \quad M_{zz1} \quad F_{z2} \quad M_{y2} \quad M_{z2} \quad M_{zz2}\right]^T$$

$$\{d\} = \left[u_1 \quad \dot{u}_1 \quad \phi_1 \quad \dot{\phi}_1 \quad u_2 \quad \dot{u}_2 \quad \phi_2 \quad \dot{\phi}_2\right]^T$$

The stiffness and geometric matrices are given in Table 5.1 in which $(M_{x_{1-2}} = M_{x_1} - M_{x_2})$.

Solution Check

It is possible to immediately check the finite element model by performing a single element solution recognising that this will not be completely accurate. The exact solution is obtained from eqn 3.7.

Consider a single element solution for a beam under uniform bending moment only. The applied loading is given by $M_{x1} = -M_{x2} = M_{cr}$. Due to symmetry $\dot{u}_2 = -\dot{u}_1$ and $\dot{\phi}_2 = -\dot{\phi}_1$. All other joint displacements $(u_1 \quad \phi_1 \quad u_2 \quad \phi_2)$ are zero. Applying these conditions to the stiffness expression leads to

$$\begin{vmatrix} \dfrac{2EI_y}{L} & \dfrac{L}{6} M_{cr} \\[3mm] \dfrac{L}{6} M_{cr} & \left(\dfrac{5}{30}GJL + \dfrac{2EI_w}{L}\right) \end{vmatrix} = 0 \qquad 5.22$$

from which the critical bending moment is given by

$$M_{cr} = \pm \frac{3}{L}\sqrt{EI_y GJ\left(\frac{4}{3} + \frac{16EI_w}{GJL^2}\right)} \qquad 5.23$$

The exact solution is $M_{cr} = \pm\dfrac{\pi}{L}\sqrt{EI_y GJ\left(1 + \dfrac{\pi^2 EI_w}{GJL^2}\right)}$

It will be seen in Section 5.5.3 that the two element solution is almost exact.

It can be observed that, in the case of a bi-symmetric I beam, two eigen-values exist which are equal in magnitude but of opposite sign. This reflects the fact that the geometric stiffness matrix is indefinite. In the case of a singly symmetric I beam, the positive and negative critical moments also differ in magnitude.

$$[k_f] = \frac{1}{L^3}
\begin{bmatrix}
(12EI_y) & & & & & & & \\
(-6EI_yL) & (4EI_yL^2) & & & \text{Symmetric} & & & \\
0 & 0 & \begin{pmatrix}1.2GJL^2\\+12EI_w\end{pmatrix} & & & & & \\
0 & 0 & \begin{pmatrix}-0.1GJL^3\\-6EI_wL\end{pmatrix} & \begin{pmatrix}0.1\dot{3}GJL^4\\+4EI_wL^2\end{pmatrix} & & & & \\
(-12EI_y) & (6EI_yL) & 0 & 0 & (12EI_y) & & & \\
(-6EI_yL) & (2EI_yL^2) & 0 & 0 & (6EI_yL) & (4EI_yL^2) & & \\
0 & 0 & \begin{pmatrix}-1.2GJL^2\\-12EI_w\end{pmatrix} & \begin{pmatrix}0.1GJL^3\\+6EI_wL\end{pmatrix} & 0 & 0 & \begin{pmatrix}1.2GJL^2\\+12EI_w\end{pmatrix} & \\
0 & 0 & \begin{pmatrix}-0.1GJL^3\\-6EI_wL\end{pmatrix} & \begin{pmatrix}-0.03GJL^4\\+2EI_wL^2\end{pmatrix} & 0 & 0 & \begin{pmatrix}0.1GJL^3\\+6EI_wL\end{pmatrix} & \begin{pmatrix}0.1\dot{3}GJL^4\\+4EI_wL^2\end{pmatrix}
\end{bmatrix}$$

$$[k_s] = \frac{1}{60}
\begin{bmatrix}
\left(\dfrac{72F_s}{L}\right) & & & & & & & \\
(-6F_s) & (8F_sL) & & & \text{Symmetric} & & & \\
\begin{pmatrix}\frac{36M_{x1-2}}{L}\\+3F_{y1}\\+33F_{y2}\end{pmatrix} & \begin{pmatrix}-33M_{x1-2}\\-6F_{y1}L\\-27F_{y2}L\end{pmatrix} & \left(\dfrac{72F_sI_p}{AL}\right) & & & & & \\
\begin{pmatrix}-3M_{x1-2}\\-3F_{y2}L\end{pmatrix} & \begin{pmatrix}4M_{x1-2}L\\+4F_{y1}L^2\\+3F_{y2}L^2\end{pmatrix} & \left(-\dfrac{6F_sI_p}{A}\right) & \left(\dfrac{8F_sI_pL}{A}\right) & & & & \\
\left(-\dfrac{72F_s}{L}\right) & (6F_s) & \begin{pmatrix}-\frac{36M_{x1-2}}{L}\\-3F_{y1}\\-33F_{y2}\end{pmatrix} & \begin{pmatrix}3M_{x1-2}\\+3F_{y2}L\end{pmatrix} & \left(\dfrac{72F_s}{L}\right) & & & \\
(-6F_s) & (-2F_sL) & \begin{pmatrix}-3M_{x1-2}\\+3F_{y1}L\\-6F_{y2}L\end{pmatrix} & \begin{pmatrix}-M_{x1-2}L\\-F_{y1}L^2\end{pmatrix} & (6F_s) & (8F_sL) & & \\
\begin{pmatrix}-\frac{36M_{x1-2}}{L}\\-33F_{y1}\\-3F_{y2}\end{pmatrix} & \begin{pmatrix}3M_{x1-2}\\+6F_{y1}L\\-3F_{y2}L\end{pmatrix} & \left(-\dfrac{72F_sI_p}{AL}\right) & \left(\dfrac{6F_sI_p}{A}\right) & \begin{pmatrix}\frac{36M_{x1-2}}{L}\\+33F_{y1}\\+3F_{y2}\end{pmatrix} & \begin{pmatrix}33M_{x1-2}\\+27F_{y1}L\\+6F_{y2}L\end{pmatrix} & \left(\dfrac{72F_sI_p}{AL}\right) & \\
\begin{pmatrix}-3M_{x1-2}\\-3F_{y1}L\end{pmatrix} & \begin{pmatrix}-M_{x1-2}L\\-F_{y2}L^2\end{pmatrix} & \left(-\dfrac{6F_sI_p}{A}\right) & \left(-\dfrac{2F_sI_pL}{A}\right) & \begin{pmatrix}2M_{x1-2}\\+3F_{y1}L\end{pmatrix} & \begin{pmatrix}4M_{x1-2}L\\+3F_{y1}L^2\\+F_{y2}L^2\end{pmatrix} & \left(\dfrac{6F_sI_p}{A}\right) & \left(\dfrac{8F_sI_pL}{A}\right)
\end{bmatrix}$$

**Table 5.1 Element stiffness and geometric matrices for
lateral-torsional buckling.**

5.5.3 Lateral-torsional Buckling Problem Solved by Jacobi Iteration Using Additional Elements

The Jacobi technique is used to provide numerical solutions of the lateral-torsional buckling problem of Fig 5.6. The results given in Table 5.2 were obtained by shifting the eigen-values positively using eqn 5.18. The following numerical parameters were used in this study: $E, G, A, I_y, J, I_w = 1$, $M_{x1} = 1$, $M_{x2} = -1$, $M_{x1-2} = 2$, $F_{z1} = P_i = -F_{z2} = 0$, $L = 10$.

The exact solutions are given by

$$M_{cr(N)} = \pm \frac{\pi}{(L/N)} \sqrt{(EI_y)(GJ)\left(1 + \frac{\pi^2 EI_w}{GJ(L/N)^2}\right)}$$ where $M_{cr(N)}$ = critical bending moment for

N half waves.

Again, observation of the Table 5.2 results shows that some eigen-values are positive and others are negative, reflecting the indefinite nature of the system.

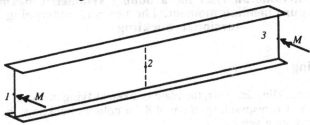

Fig 5.6 Two element lateral beam buckling problem. The beam is under uniform bending moment only.

Mode	1	2	3	4	5	6	7	8
Single Element Solution 2 Freedoms								
Eigen-values	-0.3661	0.3661						
Exact	-0.3293	0.3293						
Eigen-vectors $\dot{u}_1$	0.9504	0.9504						
$\dot{\phi}_1$	-0.3111	0.3111						
Single Element Solution 4 Freedoms								
Eigen-values	-0.3661	0.3661	-0.9798	0.97980				
Exact	-0.3293	0.3293	-0.7421	0.7421				
Eigen-vectors $\dot{u}_1$	0.6720	-0.6720	0.6030	0.6030				
$\dot{\phi}_1$	-0.2200	-0.2200	-0.3693	0.3693				
$\dot{u}_2$	-0.6720	0.6720	0.6030	0.6030				
$\dot{\phi}_2$	-0.2200	0.2200	-0.3693	0.3693				

Two Element Solution 8 Freedoms								
Eigen-values	-0.3305	0.3305	-0.8431	0.8431	-1.7158	1.7158	-2.8512	2.8512
Exact	-0.3293	0.3293	-0.7421	0.7421	-1.2951	1.2951	-2.0181	2.0181
Eigen-vectors								
$\dot{u}_1$	-0.2731	-0.2731	-0.5133	-0.5133	0.5243	-0.5243	0.4435	0.4435
$\dot{\phi}_1$	0.0816	-0.0816	0.2937	-0.2937	-0.3934	-0.3934	-0.3691	0.3691
u_2	0.8755	0.8755	0.0288	0.0288	0.2771	-0.2771	0.0125	0.0125
$\dot{u}_2$	-0.0018	-0.0018	0.5012	0.5012	0.0011	-0.0011	0.4420	0.4420
ϕ_2	-0.2632	0.2632	-0.0226	0.0226	-0.2078	--0.2078	-0.0290	0.0290
$\dot{\phi}_2$	0.0038	-0.0038	-0.2853	0.2853	0.0026	0.0026	-0.3712	0.3712
$\dot{u}_3$	0.2759	0.2759	-0.4893	-0.4893	-0.5368	0.5368	0.4400	0.4400
$\dot{\phi}_3$	-0.0833	0.0833	0.2773	-0.2773	0.4027	0.4027	-0.3732	0.3732

Table 5.2 Eigen-value analysis for a doubly symmetric beam under pure bending moment. The beam is undergoing lateral -torsional buckling.

5.6 Plate Buckling

In plate bending and buckling analysis, the plate is taken as lying in the x, y plane; the z axis is normal to the plate. Corresponding to eqn 4.8 for column buckling, the large deflection expressions for strains are given by

$$\varepsilon_{xx} = \frac{\partial u}{\partial x} - z\frac{\partial^2 w}{\partial x^2} + \frac{1}{2}\left(\frac{\partial w}{\partial x}\right)^2 \qquad 5.24a$$

$$\varepsilon_{yy} = \frac{\partial v}{\partial x} - z\frac{\partial^2 w}{\partial y^2} + \frac{1}{2}\left(\frac{\partial w}{\partial y}\right)^2 \qquad 5.24b$$

$$\varepsilon_{xy} = \frac{\partial u}{\partial y} + \frac{\partial u}{\partial y} - 2z\frac{\partial^2 w}{\partial x\partial y} + \frac{\partial w}{\partial x}\frac{\partial w}{\partial y} \qquad 5.24c$$

and the strain energy by

$$U = 0.5D\int_A\left[\left(\frac{\partial^2 w}{\partial x^2} + \frac{\partial^2 w}{\partial y^2}\right)^2 + 2v\frac{\partial^2 w}{\partial x^2}\frac{\partial^2 w}{\partial y^2} + 2(1-v)\left(\frac{\partial^2 w}{\partial x\partial y}\right)^2\right]dA$$

$$+ 0.5\int_A \sigma_x t\left(\frac{\partial w}{\partial x}\right)^2 dA + 0.5\int_A \sigma_y t\left(\frac{\partial w}{\partial y}\right)^2 dA \qquad 5.25$$

$$+ 0.5\int_A \sigma_{xy} t\left(\frac{\partial w}{\partial x}\right)\left(\frac{\partial w}{\partial y}\right)dA$$

where

$D = Et^3/12(1 - v^2)$ = flexural rigidity,

t = plate thickness

v = Poisson's ratio

$\sigma_x t = N_x$ = direct force per unit width acting in the plane of the plate in the x direction

$\sigma_y t = N_y$ = direct force per unit width acting in the plane of the plate in the y direction

$\sigma_{xy} t = N_{xy}$ = shear force per unit width acting in the plane of the plate in the x direction

The development of the finite expressions follows generally similar procedures to those used to develop finite element expressions for column and lateral-torsional buckling problems. The displacement functions are given in the form

$$w = Nd \qquad\qquad 5.26$$

After differentiation and substitution into eqn 5.25 the flexural strain energy is given by

$$U = \tfrac{1}{2} d^T k_f d + \tfrac{1}{2} d^T k_{gx} d + \tfrac{1}{2} d^T k_{gy} d + \tfrac{1}{2} d^T k_{gxy} d \qquad\qquad 5.27$$

Typically

$$k_{gx} = \int_A \sigma_x t \left[\frac{\partial N}{\partial x} \right]^T \left[\frac{\partial N}{\partial x} \right] dA \qquad\qquad 5.28$$

For many plate problems, the computation of the buckling load is not of any particular significance as plates have considerable reserves of strength in the post-buckling range, particularly when loaded in shear. The reader is referred to the many specialist publications which deal more specifically with the topic of plate post-buckling behaviour. Material non-linearities commonly play a significant role in plate post-buckling behaviour.

6   MATRIX BASICS

In developing the general characteristics of eigen-analysis some basic properties of symmetric matrices become significant. These basic properties are now introduced.

6.1   Symmetric Matrices

Symmetric matrices play a central role in structural vibration and stability theory; minimum potential energy formulations always lead to symmetric stiffness and geometric matrices. Moreover, stress analysis is not the only field of mathematical physics in which symmetric systems of equations arise. As a result, considerable effort has been expended in determining the properties of and developing methods to extract symmetric system eigen-pairs (eigen-values and eigen-vectors). Most computer algorithms for extracting eigen-pairs work only with symmetric matrices.

6.1.1 Definition of a Symmetric Matrix

A symmetric matrix, K, is one for which the elements satisfy the condition $k_{ij} = k_{ji}$, $i \neq j$. For example, if

$$K = \begin{bmatrix} k_{11} & k_{12} & k_{13} & k_{14} \\ k_{21} & k_{22} & k_{23} & k_{24} \\ k_{31} & k_{32} & k_{33} & k_{34} \\ k_{41} & k_{42} & k_{43} & k_{44} \end{bmatrix}$$

then $k_{12} = k_{21}$ etc.

6.1.2 Symmetric Matrices and Quadratic Products

A quadratic matrix product has a matrix product having the form $U = w^T K v$, where v and w are vectors, ie, (nx1) matrices, and K has dimensions (nxn). The result, U, is a scalar, ie, a (1x1) matrix. For K symmetric, the quadratic product possess the property $U = w^T K v = v^T K w$. Engineers can view quadratic products as being tantamount to strain energy since, if K is a symmetric structural stiffness matrix, and $v = w = D =$ displacements, then $U = 0.5 D^T K D$ is the usual expression for strain energy.

Proof $(w^T K v = v^T K w)$
Using the summation convention of repeated subscripts the left quadratic product, U_1, is given by

$$U_l = w^T K v = w_i k_{ij} v_j \tag{6.1}$$

Similarly the right quadratic product, U_2, is given by

$$U_r = v^T K w = v_i k_{ij} w_j \tag{6.2}$$

Since i and j are dummy subscripts, they may be interchanged so that

$$U_r = v^T K w = v_j k_{ji} w_i \tag{6.3}$$

However, the matrix K is symmetric, so that $k_{ij} = k_{ji}$ which leads to the result $U = U_l = U_r$.

Example 6.1

Consider the quadratic product $U = \begin{bmatrix} 4 & 5 & 6 \end{bmatrix} \begin{bmatrix} 3 & 4 & 5 \\ 4 & 2 & 1 \\ 5 & 1 & 3 \end{bmatrix} \begin{bmatrix} 1 \\ 2 \\ 3 \end{bmatrix} = 255$

Also $U = \begin{bmatrix} 1 & 2 & 3 \end{bmatrix} \begin{bmatrix} 3 & 4 & 5 \\ 4 & 2 & 1 \\ 5 & 1 & 3 \end{bmatrix} \begin{bmatrix} 4 \\ 5 \\ 6 \end{bmatrix} = 255$

6.2   Eigen-problems

The most commonly discussed eigen-problems has the standard form

$$(K - \lambda I)D = 0 \qquad\qquad 6.4$$

where λ = eigen values, I = unit matrix, D = vector of unknowns (the displacements in first order stability and vibration analysis). Such systems always have a trivial solution of $D^T = \langle 0 \quad 0 \quad 0 \quad . . \quad 0 \rangle$ but there exists other solutions involving non-null displacement matrices for particular values of λ, the eigen-values.

6.2.1 Orthogonality Property for the Eigen-vectors of Symmetric Matrices

The eigen-vectors of a symmetric matrix are orthogonal. If they are also normalised and designated by e_1, e_2, e_3, ...the orthogonality property implies that

$$e_i^T e_j = 0 \qquad i \neq j \qquad\qquad 6.5(a)$$
$$e_i^T e_j = 1 \qquad i = j \qquad\qquad 6.5(b)$$

or

$$e_i^T e_j = \delta_{ij} \qquad\qquad 6.5(c)$$

where δ_{ij} = Kronecker delta

Proof

Let two of the eigen-pairs be (λ_i, e_i) and (λ_j, e_j). It follows that, for the i*th* eigen-pair,

$$Ke_i - \lambda_i I e_i = 0 \qquad\qquad 6.6$$

Premultiply eqn 6.6 by e_j^T which leads to

$$e_j^T Ke_i - \lambda_i e_j^T I e_i = 0 \qquad\qquad 6.7$$

Because K is symmetric, it follows from the quadratic product property of symmetric matrices that eqn 6.7can be rewritten as

$$e_i^T Ke_j - \lambda_i e_i^T I e_j = 0 \qquad\qquad 6.8$$

Now consider the j*th* eigen-pair

$$Ke_j - \lambda_j Ie_j = 0 \qquad\qquad 6.9$$

or pre-multiplying by e_i^T

$$e_i^T Ke_j - \lambda_j e_i^T Ie_j = 0 \qquad\qquad 6.10$$

After subtracting eqn 6.10 from eqn 6.8, it follows that $(\lambda_i - \lambda_j)e_j^T e_i = 0$. Provided, $\lambda_i \neq \lambda_j$, then $e_j^T e_i = 0$, as stated in eqn 6.5(a). In the case $i = j$, the expression becomes $(\lambda_i - \lambda_i)e_i^T e_i = 0$ and, for all non-null e_i, the product $e_i^T e_i$ represents the square of the eigen-vector Euclidean norm. But the vectors are normalised such that $e_i^T e_i = 1$.

Example 6.2

For a prismatic (constant section) beam element having nodal displacements v_1, θ_1, v_2, θ_2 the stiffness matrix is given by

$$k_f = \begin{bmatrix} \dfrac{12EI}{L^3} & symmetric & & \\ \dfrac{-6EI}{L^2} & \dfrac{4EI}{L} & & \\ \dfrac{-12EI}{L^3} & \dfrac{6EI}{L^2} & \dfrac{12EI}{L^3} & \\ \dfrac{-6EI}{L^2} & \dfrac{2EI}{L} & \dfrac{6EI}{L^2} & \dfrac{4EI}{L} \end{bmatrix} = \begin{bmatrix} 12 & symmetric & & \\ -6 & 4 & & \\ -12 & 6 & 12 & \\ -6 & 2 & 6 & 4 \end{bmatrix}$$

if $EI = 1$, $L = 1$.

An eigen-analysis produces the following normalised eigen-pairs

Mode 1	Mode 2	Mode 3	Mode 4	
$\begin{bmatrix} 0.63246 \\ 0.31623 \\ -0.63246 \\ 0.31623 \end{bmatrix}$	$\begin{bmatrix} -0.31623 \\ 0.63246 \\ 0.31623 \\ 0.63246 \end{bmatrix}$	$\begin{bmatrix} 0.70711 \\ 0.00000 \\ 0.70711 \\ 0.00000 \end{bmatrix}$	$\begin{bmatrix} 0.00000 \\ -0.70711 \\ 0.00000 \\ 0.70711 \end{bmatrix}$	Eigen-vectors
30	0	0	2	Eigen-values

It is easy to verify that these vectors are orthogonal. The displacement patterns corresponding to the eigen-vectors are illustrated below.

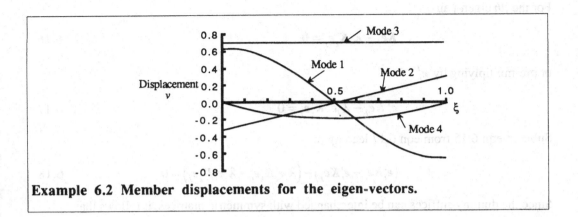

Example 6.2 Member displacements for the eigen-vectors.

6.2.2 Orthogonality Property for Generalised, Symmetric Eigen-problems

Theorem 1

Eigen-problems arising in the analysis of structural stability and vibration take the more general form

$$(K - \lambda K_g)D = 0 \qquad\qquad 6.11$$

where K = symmetric stiffness matrix of dimensions $(n \times n)$, K_g = symmetric geometric matrix, D = vector of displacement components. Corresponding to the statements, eqn 6.7(c), it is also possible, for suitably scaled e_i and e_j, to write

$$e_j^T K_g e_i = \delta_{ij} \qquad\qquad 6.12$$

Proof

Let two of the eigen-pairs be (λ_i, e_i) and (λ_j, e_j). It follows that, for the *ith* eigen-pair,

$$Ke_i - \lambda_i K_g e_i = 0 \qquad\qquad 6.13$$

Premultiply eqn 6.14 by e_j^T which leads to

$$e_j^T Ke_i - \lambda_i e_j^T K_g e_i = 0 \qquad\qquad 6.14$$

Because K and K_g are symmetric, it follows from the quadratic product discussed in Section 6.5.2 that eqn 6.13 can be rewritten as

$$e_i^T Ke_j - \lambda_i e_i^T K_g e_j = 0 \qquad\qquad 6.15$$

For the jth eigen-pair

$$Ke_j - \lambda_j K_g e_j = 0 \qquad\qquad 6.16$$

or pre-multiplying by e_i^T

$$e_i^T Ke_j - \lambda_j e_i^T K_g e_j = 0 \qquad\qquad 6.17$$

Subtract eqn 6.15 from eqn 6.17 leading to

$$\left(e_j^t Ke_i - e_i^t Ke_j\right) - \left(\lambda_i e_j^t K_g e_i - \lambda_j e_i^t K_g e_j\right) = 0 \qquad\qquad 6.18$$

Since the dummy suffices can be interchanged with symmetric matrices, it follows that

$$\left(\lambda_i - \lambda_j\right)e_j^t K_g e_i = 0 \qquad\qquad 6.19$$

Hence, if $i \neq j$ and $\lambda_i \neq \lambda_j$, then

$$e_j^T K_g e_i = 0 \qquad i \neq j \qquad\qquad 6.20$$

It also follows, since $e_j^t Ke_i = \lambda_i e_j^t K_g e_i$, that

$$e_j^T Ke_i = 0 \qquad i \neq j \qquad\qquad 6.21$$

For the case $i = j$, by appropriate scaling of e_i, it also follows that

$$e_i^T K_g e_j = 1 \qquad\qquad 6.22$$

and that

$$e_i^T Ke_j = \lambda_i \qquad\qquad 6.23$$

6.2.3 Definiteness of Symmetric Matrices

Arising out of minimum potential energy analysis, are quadratic products of the type $D^T KD$.

For a stiffness matrix, irrespective of how D is chosen, the quadratic product $D^T KD$ is always positive if the structure is restrained against rigid body motion (Example 6.2); from physical considerations alone, strain energy is positive. However, if rigid body displacements are not restrained, then $D^T KD$ can be zero, for choices of D involving rigid body movement only, and K is said to be positive semi-definite.

Buckling problems arise in the form $(K - \lambda K_g)D = 0$. K is either positive definite or positive semi-definite but, depending on the problem, K_g may be positive or negative definite, positive or negative semi-definite, indefinite or semi-indefinite. For an Euler column problem the K_g matrix is positive or negative definite depending on the sign convention chosen for axial load. On the other hand, the K_g matrix given in Table 5.1 for beam lateral-torsional buckling is indefinite for the case of uniform bending moment; for doubly symmetric I beams, there exists equal but oppositely signed critical moments. If the beam is unsymmetrical about the major axis even the numerical values of the positive and negative critical moments differ.

The fact that K_g can form quadratic products having varying sign makes buckling problems of the eigen type more difficult to deal with than vibration problems which have positive definite mass matrices; vibration frequencies can only be positive.

6.3 Rayleigh Quotients

Rayleigh quotients are scalars defined by the expression

$$\rho = \frac{D^T K D}{D^T K_g D} \qquad\qquad 6.24$$

and are associated with the eigen-problem $(K - \lambda K_g)D = 0$ or $D^T K D - D^T \lambda K_g D = 0$. In a positive definite system the minimum eigen-value (critical load) is related to the Rayleigh quotient by

$$\lambda_{min} = min\left[\frac{D^T K D}{D^T K_g D}\right] \qquad\qquad 6.25$$

If D is scaled so that $D^T K D = 1$ then λ_{min} is computed when $D^T K_g D$ is maximum. More generally, if $D = e_i$, then $\rho = \frac{e_i^T K e_i}{e_i^T K_g e_i} = \lambda_i$. Many approximate buckling and vibration analyses rely upon evaluations of Rayleigh quotients for trial choices of displacement functions ($v = Nd$ in the case of finite element or Rayleigh-Ritz analysis) which appear to be intuitively reasonable. It is usual to choose functions which meet the kinematical boundary conditions.

Of all the choices made which satisfy the kinematical boundary conditions, the one which gives the lowest Rayleigh quotient provides the best estimate of the minimum critical load. The choice of shapes close to other critical modes will give upper bound estimates of the higher critical loads.

If the geometric matrix, K_g, is positive definite or negative definite then all Rayleigh Quotients are, respectively, positive or negative by definition. If the geometric matrix is indefinite then there exists both positive and negative Rayleigh Quotients; the sign of the Rayleigh Quotient depends on the choice of deformed shape.

6.3.1 Minimum Property of Rayleigh Quotients for Symmetric, Positive Definite Systems

Theorem 2

Suppose that the eigen-values of a symmetric, positive definite system

$$(K - \lambda K_g)D = 0 \qquad 6.26$$

are written in ascending order as $\lambda_1, \lambda_2, \ldots, \lambda_n$. Let the eigen-pair corresponding to the minimum eigen-value be (λ_1, e_1). There exists no Rayleigh Quotient less than λ_1.

Proof

The eigen-vectors always form a basis of a vector space; see Section 6.3. Let an arbitrary vector D be given by a linear combination of the eigen-vectors as

$$D = \sum_{i=1}^{n} \alpha_i e_i \qquad 6.27$$

The Rayleigh Quotient can be expressed in terms of the eigen-vectors as

$$\rho = \frac{D^T K D}{D^T K_g D} = \frac{\left(\sum_{j=1}^{n} \alpha_j e_j\right)^T K \sum_{i=1}^{n} \alpha_i e_i}{\left(\sum_{j=1}^{n} \alpha_j e_j\right)^T K_g \sum_{i=1}^{n} \alpha_i e_i} = \frac{\sum_{i=1}^{n} \alpha_i^2 e_i^T K e_i}{\sum_{i=1}^{n} \alpha_i^2 e_i^T K_g e_i} \qquad 6.28$$

The expansion uses the orthogonality property of the eigen-vectors to eliminate the cross-product terms. Eqn 6.28 can be expanded in the form

$$\rho = \frac{\alpha_1^2 e_1^T K e_1 + \sum_{i=2}^{n} \alpha_i^2 e_i^T K e_i}{\alpha_1^2 e_1^T K_g e_1 + \sum_{i=2}^{n} \alpha_i^2 e_i^T K_g e_i} = \lambda_1 \left\{ \frac{\alpha_1^2 e_1^T K_g e_1 + \sum_{i=2}^{n} \alpha_i^2 \frac{\lambda_i}{\lambda_1} e_i^T K_g e_i}{\alpha_1^2 e_1^T K_g e_1 + \sum_{i=2}^{n} \alpha_i^2 e_i^T K_g e_i} \right\} \qquad 6.29$$

Given that K_g is positive definite, all terms $e_i^T K_g e_i$ are positive. Further, since λ_1 is the minimum eigen-value, all terms $\left. \dfrac{\lambda_i}{\lambda_1} \geq 1 \right|_{i=2}^{n}$. It follows that the term in curly brackets is always greater than or equal to 1 for any choice of the α values.

6.3.2 Maximum Property of Rayleigh Quotients for Positive Definite Systems

Theorem 3

Let the eigen-pair corresponding to the maximum eigen-value be (λ_n, e_n). There exists no Rayleigh Quotient greater than λ_{max}.

Proof

The proof is trivial. Since the maximum eigen-pair satisfy $(K - \lambda_n K_g)e_n = 0$ then dividing by λ_n leads to $\left(K_g - \dfrac{1}{\lambda_n} K \right) e_n = (K_g - \beta_1 K)e_n = 0$ where β_1 is the smallest eigen-value of the system $(K_g - \beta K)v = 0$. The proof is automatically established from Theorem 2. However, if K_g is indefinite, then the upper bound may be infinite.

6.3.3 Range of Rayleigh Quotients

Because the Rayleigh Quotients are bounded above and below, it follows that all such quotients are constrained to a range of values as illustrated in Fig 6.2(a).

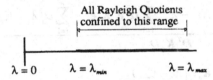

(a) K and K_g are both positive definite

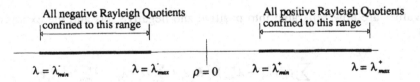

(b) K is positive definite and K_g is negative definite

Fig 6.1: Range of Rayleigh Quotient values for:

(a) positive definite systems of the form $(K - \lambda K_g)D = 0$; K and K_g are positive definite, all eigen-values and Rayleigh Quotients are positive,

(b) indefinite systems of the form $(K - \lambda K_g)D = 0$; K is positive definite and K_g is indefinite, eigen-values and Rayleigh Quotients may be positive or negative.

6.3.4   Separation Property for Indefinite Systems

Theorem 4

With K_g indefinite, the Rayleigh Quotients separate into positive and negative ranges; see Fig 6.1(b). No Rayleigh Quotients lie in the range between λ_{max}^- and λ_{min}^+. It is also true that no Rayleigh Quotients lie, respectively, below nor above λ_{min}^- and λ_{max}^+.

Proof

Let an arbitrary vector D be given by a linear combination of the eigen-vetors as

$$D = \sum_{i=1}^{n} \alpha_i e_i \qquad 6.30$$

Consider, first, the minimum positive eigen-value and subtract it from any positive Rayleigh Quotients. The assumption is made that, in some instances, the eigen-values are positive and, in others, negative depending on the choice displacement vector.

$$\rho - \lambda_{min}^+ = \frac{D^T K D}{D^T K_g D} - \lambda_{min}^+ = \frac{D^T K D - \lambda_{min}^+ D^T K_g D}{D^T K_g D}$$

$$= \frac{\sum_{i=1}^{n} \alpha_i^2 e_i^T K e_i - \lambda_{min}^+ \sum_{i=1}^{n} \alpha_i^2 e_i^T K_g e_i}{D^T K_g D} = \frac{\sum_{i=1}^{n} \alpha_i^2 \lambda_i e_i^T K_g e_i - \lambda_{min}^+ \sum_{i=1}^{n} \alpha_i^2 e_i^T K_g e_i}{D^T K_g D}$$

$$= \frac{\sum_{i=1}^{n} \alpha_i^2 e_i^T K_g e_i (\lambda_i - \lambda_{min}^+)}{D^T K_g D}$$

The eigen-values are now separated into positive and negative sets and the expression is rewritten as

$$\rho - \lambda_{min}^+ = \frac{\sum_{i=1}^{n} \alpha_i^2 (\lambda_i^+ - \lambda_{min}^+) e_i^T K_g e_i + \sum_{i=1}^{n} \alpha_i^2 (\lambda_i^- - \lambda_{min}^+) e_i^T K_g e_i}{D^T K_g D} \geq 0 \qquad 6.31$$

where the first summation in the numerator is only for positive eigen-values and the second summation is only for negative eigen-values.

Consider the following.

i)   In buckling problems, the sign of a Rayleigh Quotient is always determined by the sign of the denominator. With an indefinite matrix, K_g, a positive Rayleigh Quotient can only arise if D is selected such that $D^T K_g D$ is positive.

ii)   The first summation in the numerator is taken only over positive eigen-values, λ_i^+, which are associated with positive values of $e_i^T K_g e_i$. Hence, all terms in the expression $\alpha_i^2 (\lambda_i^+ - \lambda_{min}^+) e_i^T K_g e_i$ are positive.

iii) The second summation in the numerator is taken only over negative eigen-values, λ_i^-, which are associated with negative values of $e_i^T K_g e_i$. Thus, while $(\lambda_i^- - \lambda_{min}^+)$ is negative the corresponding $e_i^T K_g e_i$ is negative and $\alpha_i^2 (\lambda_i^- - \lambda_{min}^+) e_i^T K_g e_i$ is positive.

Since all terms in the numerator and denominator are positive it follows that $\rho - \lambda_{min}^+ \geq 0$.

By a similar approach it is possible to show that, over the range of negative Rayleigh Quotients, $\rho - \lambda_{max}^- \leq 0$.

By inverting the problem, as was done for positive definite systems, it can be shown that $\rho - \lambda_{max}^+ \leq 0$ and $\rho - \lambda_{max}^- \geq 0$. A full statement is $\lambda_{max}^- \leq \rho^- \leq \lambda_{min}^-$ and $\lambda_{min}^+ \leq \rho^+ \leq \lambda_{max}^+$.

6.4 Increasing Structure Stiffness - Structure Size Remains Constant

From a mathematical point of view, the simplest way of increasing buckling capacity is by increasing the stiffness of individual members, eg, by increasing the flexural stiffness of members subject to Euler buckling. It is assumed that K is a positive-definite parent stiffness matrix which is augmented by a positive semi-definite matrix ΔK. The reason for saying that ΔK is positive-semi-definite is that, in a structure, only some members may be stiffened. ΔK will have non-zero terms in rows and columns affected by the stiffening and null rows and columns corresponding to displacement components not affected by the stiffening.

The matrix K_g is assumed to remain constant during the change ΔK.

Theorem 5

Suppose that a system $(K - \lambda K_g)D = 0$ has minimum positive and negative eigen-values λ_{min}^+ and λ_{max}^-. An increase in structure stiffness matrix given is given by $K + \Delta K$ results in no Rayleigh Quotients of the system $(K + \Delta K - \lambda K_g)D = 0$ lying in the range λ_{min}^+ to λ_{max}^-.

Proof

The amended Rayleigh Quotients are given by

$$\rho + \Delta\rho = \frac{D^T(K + \Delta K)D}{D^T K_g D}$$
6.32

Consider first only positive Rayleigh Quotients.

$$\rho + \Delta\rho = \frac{D^T KD}{D^T K_g D} + \frac{D^T \Delta KD}{D^T K_g D}$$
6.33

Since the first term has a minimum of λ_{min}^+ and the second is zero or positive, it follows that $\Delta\rho$ is positive and that $(\rho + \Delta\rho)^+ \geq \lambda_{min}^+$. By a similar argument, it is possible to consider only negative Rayleigh Quotients and to argue that $(\rho + \Delta\rho)^- \leq \lambda_{max}^-$.

Application

The type of practical situation to which Theorem 5 applies is illustrated in Fig 6.2 which is a statically determinate truss in which one member has been stiffened.

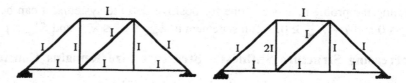

Fig 6.2 Statically determinate truss in which one member is stiffened - Theorem 5 shows that the critical load factors cannot be reduced by such action.

In statically determinate structures the internal stress distributions and therefore the k_g matrices are not affected by the element stiffness characteristics. In a statically determinate truss, for example, the P_i are not affected by member EA/L values; see below. Thus any changes in member EA, which accompany changes in EI, leave k_g unaltered.

$$k_g = \frac{P_i}{30L}\begin{bmatrix} 36 & & symmetric & \\ -3L & 4L^2 & & \\ -36 & 3L & 36 & \\ -3L & -L^2 & 3L & 4L^2 \end{bmatrix}$$

Reference is also made to Example 5.2 involving a statically indeterminate structure. In this example, an increase in the spring stiffness, k_2, causes a reduction in critical load because the P_i in the column increases without increase in its buckling resistance. In more general terms, it is the internal stress resultants induced in members by the external loading which contributes to instability rather than the external loading itself; see also how the plate buckling problems are formulated in Section 5.6. The term σ_x in the expression

$k_{gx} = \int_A \sigma_x t \left[\frac{\partial N}{\partial x} \right]' \left[\frac{\partial N}{\partial x} \right] dA$ is the internal stress at a specific point in the plate. Irrespective of

the buckling type, the internal stress distribution can be affected by changes to element stiffness characteristics. For this reason, in the proof of Theorem 5, matrices k_g and K_g are assumed invariable under change in ΔK.

6.5   Nett Positive Definiteness of a Loaded Structure

Theorem 6

It is possible to make observations about the definiteness characteristics of structural systems where the K and K_g matrices are considered separately as above. It is invariably true that K remains positive definite, ie, it always forms a positive quadratic product, $D^t K D$, with displacements representing twice the strain energy. On the other hand, the K_g matrix may be positive or negative definite or indefinite depending on the nature of the instability phenomenon.

It is also possible to make observations of quadratic products formed with $K - \lambda K_g$, viz, $D^t (K - \lambda K_g) D$, in situations where $\lambda_{max}^- \leq \lambda \leq \lambda_{min}^+$. With the load factor constrained in this manner, $K - \lambda K_g$ is always positive definite.

Proof

Let

$$X = D^T (K - \lambda K_g) D = D^T K_g D \left[\frac{D^T K D}{D^T K_g D} - \lambda \right] \qquad 6.34$$

where $\lambda_{max}^- \leq \lambda \leq \lambda_{min}^+$. The quantity X measures the definiteness characteristics of the system $(K - \lambda K_g) D = 0$.

Restrict attention, initially, to all choices of D which cause $D^T K_g D$ to remain positive. X has sign determined by $X = +E[\lambda_{min}^+ + \varepsilon - \lambda]$ where $E = D^T K_g D$ and $\varepsilon = \frac{D^T K_g D}{D^T K D}$ are both positive constants and $\lambda \leq \lambda_{min}^+$. Since the quantity in square brackets remains positive, X remains positive for all choices of D which cause $D^T K_g D$ to remain positive.

Consider the alternative which involves all choices of D which cause $D^T K_g D$ to remain negative. X has sign determined by $X = -E[\lambda_{max}^- - \varepsilon - \lambda]$ where E, ε are positive constants and $\lambda_{max}^- \leq \lambda$. Since the quantity in square brackets remains negative, X remains positive for all choices of D which cause $D^T K_g D$ to remain negative.

As a consequence, $X = D^T (K - \lambda K_g) D$ remains positive so long as $\lambda_{max}^- \leq \lambda \leq \lambda_{min}^+$.

6.6 Combining the Parent Structure with a Supplementary Structure having Nett Positive Definiteness

Theorem 7

Consider two interconnected structures subject to load patterns proportional to independent load factors, $\lambda_{(p)}$, for a parent structure, and $\lambda_{(s)}$, for a supplementary structure. It is natural to restrict the load factors of each structure to the ranges $\lambda_{(p)max}^- \leq \lambda_{(p)} \leq \lambda_{(p)min}^+$ and $\lambda_{(s)max}^- \leq \lambda_{(s)} \leq \lambda_{(s)min}^+$ which ensures that the two structures will retain nett positive definiteness under their separate loadings and support conditions. The two structures are inter-connected in a fashion that does not affect the $K_{g(p)}$ or $K_{g(s)}$ matrices.

Such inter-connection cannot reduce the critical loads of either structure.

Proof

The inter-connection of the parent and supplementary structures will introduce, from the parent structure perspective, new displacement components. Some displacement components, D_2, will be common to the two structures. The displacements are characterised as belonging to one of the classes:

D_1, parent structure only displacements,
D_2, common displacements,
D_3, supplementary structure only displacements.

The displacements D_1 are lowest numbered and D_3 highest numbered. Common displacements, D_2, have intermediate numbers. Such ordering is always possible. The K and K_g matrices can be expanded and partitioned to correspond to this ordering.

The parent structural system is written in the form

$$\left[\begin{bmatrix} K_{(p)11} & K_{(p)12} & 0 \\ & K_{(p)22} & 0 \\ Symmetric & & 0 \end{bmatrix} - \lambda_{(p)} \begin{bmatrix} K_{(p)g11} & K_{(p)g12} & 0 \\ & K_{(p)g22} & 0 \\ Symmetric & & 0 \end{bmatrix} \right] \begin{bmatrix} D_1 \\ D_2 \\ D_3 \end{bmatrix} = \begin{bmatrix} 0 \\ 0 \\ 0 \end{bmatrix} \qquad 6.35$$

where $\begin{bmatrix} D_1 \\ D_2 \end{bmatrix} = D_{(p)}$ in the original statement of the eigen-problem for the parent structure $(K_{(p)} - \lambda_{(p)} K_{(p)g}) D_{(p)} = 0$.

For the supplementary structure, the system $(K_{(s)} - \lambda_{(s)} K_{(s)g}) D_{(s)} = 0$, is rewritten as

$$\begin{bmatrix} 0 & 0 & 0 \\ & (K_{(s)22} - \lambda_{(s)} K_{(s)g22}) & (K_{(s)23} - \lambda_{(s)} K_{(s)g23}) \\ Symmetric & & (K_{(s)33} - \lambda_{(s)} K_{(s)g33}) \end{bmatrix} \begin{bmatrix} D_1 \\ D_2 \\ D_3 \end{bmatrix}$$

$$= \begin{bmatrix} 0 & 0 & 0 \\ & K'_{(s)22} & K'_{(s)23} \\ Symmetric & & K'_{(s)33} \end{bmatrix} \begin{bmatrix} D_1 \\ D_2 \\ D_3 \end{bmatrix} = \begin{bmatrix} 0 \\ 0 \\ 0 \end{bmatrix} \qquad 6.36$$

The stiffness matrices in eqn 6.36 are always positive definite, provided $\lambda^-_{(s)max} \leq \lambda_{(s)} \leq \lambda^+_{(s)min}$. Adding eqn 6.36 to eqn 6.35 leads to

$$\begin{bmatrix} \begin{bmatrix} K_{(p)11} & K_{(p)12} & 0 \\ & K_{(p)22} + K'_{(s)22} & K'_{(s)23} \\ Symmetric & & K'_{(s)33} \end{bmatrix} - \lambda_{(p)} \begin{bmatrix} K_{(p)g11} & K_{(p)g12} & 0 \\ & K_{(p)g22} & 0 \\ Symmetric & & 0 \end{bmatrix} \end{bmatrix} \begin{bmatrix} D_1 \\ D_2 \\ D_3 \end{bmatrix} = \begin{bmatrix} 0 \\ 0 \\ 0 \end{bmatrix} \quad 6.37$$

For the parent structure taken in isolation, the eigen problem has the form

$$\begin{bmatrix} \begin{bmatrix} K_{(p)11} & K_{(p)12} \\ Symmetric & K_{(p)22} + K'_{(s)22} \end{bmatrix} - \lambda_{(p)} \begin{bmatrix} K_{(p)g11} & K_{(p)g12} \\ Symmetric & K_{(p)g22} \end{bmatrix} \end{bmatrix} \begin{bmatrix} D_1 \\ D_2 \end{bmatrix} = \begin{bmatrix} 0 \\ 0 \end{bmatrix} \qquad 6.38(a)$$

$$(K + \Delta K - \lambda_{(p)} K_g) D_{(p)} = 0 \qquad 6.38(b)$$

According to Theorem 5 the parent structure cannot have any of its eigen-values reduced in absolute value by connection to a structure which is, itself, stable, provided that there is no force transfer between the two structures. Put another way, provided $\lambda^-_{(s)max} \leq \lambda_{(s)} \leq \lambda^+_{(s)min}$, then no $\rho_{(p)}$ can lie in the range $\lambda^-_{(p)max} \leq \rho_{(p)} \leq \lambda^+_{(p)min}$ after the structures are inter-connected.

A similar statement can be made of the supplementary structure, ie, provided $\lambda^-_{(p)max} \leq \lambda_{(p)} \leq \lambda^+_{(p)min}$ then no $\rho_{(s)}$ can lie in the range $\lambda^-_{(s)max} \leq \rho_{(s)} \leq \lambda^+_{(s)min}$ after the structures are inter-connected. The proof of this result follows from reversing the roles of the parent and supplementary stiffness matrices.

Application

An application of Theorem 7 is illustrated in Fig 6.5 where the parent and supplementary structure are connected after the basic stress distribution is developed separately in each structure up to the basic stress level. The basic stress distribution is the one factored in the eigen-analysis. The composite structure is stable provided the load factor for the composite structure does not exceed the lesser of the minimum critical load factors of the two structures.

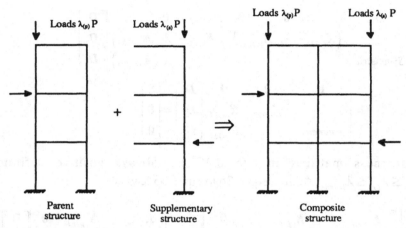

Fig 6.3   Parent structure and supplementary structure under independent load systems - both structures stable. The composite structure is stable provided the load factor for the composite structure does not exceed the lesser of the minimum critical load factors of the two structures.

6.7   Combining the Parent Structure with a Supplementary Structure having Nett Positive Definiteness; Both Structures have a Common Load Factor

Where the parent and supplementary structures have a common load factor, λ, a statement similar to that of Section 6.10 can be made. It takes the form that no Rayleigh Quotient, ρ, of the composite structure can be found in the range

$$\left(\lambda_{(p)max}^{-}, \lambda_{(s)max}^{-}\right)_{max} \leq \rho \leq \left(\lambda_{(p)min}^{+}, \lambda_{(s)min}^{+}\right)_{min} \qquad 6.39$$

provided the load factor, λ, for the composite system is constrained such that

$$\left(\lambda_{(p)max}^{-}, \lambda_{(s)max}^{-}\right)_{max} \leq \lambda \leq \left(\lambda_{(p)min}^{+}, \lambda_{(s)min}^{+}\right)_{min} \qquad 6.40$$

Previous theorems have been developed which predict the properties of a composite structure formed by the coalescence of a parent and supplementary structure. In these theorems independent, zero order analyses are carried out on the parent and supplementary structures in isolation. In the present theorem, the zero order analysis is carried out on the composite structure so that the base internal stress distribution includes the effect of any load sharing between the two structures. For trusses, the $K_{g(p)}$ and $K_{g(s)}$ are based on the P_i values determined for a zero order analysis of the composite structure. If eigen-analyses are subsequently performed on the parent and supplementary structures, taken in isolation, using the $K_{g(p)}$ and $K_{g(s)}$ determined from the base stress distributions, respectively, then changes $\Delta K_{(p)}$ and $\Delta K_{(s)}$ in $K_{(p)}$ and $K_{(s)}$, respectively, cannot reduce the critical load factors of the composite structure.

Proof

The proof follows automatically from Theorem 7.

Application

Consider a portal frame illustrated in Fig 6.4 susceptible to a known bending moment distribution and susceptible to out-of-plane (lateral-torsional buckling). The zero order analysis is performed on the complete portal frame but sub-structures may be isolated therefrom and designed to be stable in isolation - assume that this is the rafter in Fig 6.4. From the point of view of the language used in Theorem 8 the isolated rafter can be viewed as the parent structure. If the resistance to lateral-torsional buckling of the isolated rafter is increased then the resistance of the composite structure is increased.

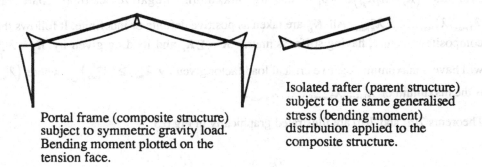

Portal frame (composite structure) subject to symmetric gravity load. Bending moment plotted on the tension face.

Isolated rafter (parent structure) subject to the same generalised stress (bending moment) distribution applied to the composite structure.

Fig 6.4: Portal frame susceptible to lateral torsional buckling.

Where the in-plane buckling of a statically indeterminate frame is under consideration, the base stress distribution is derived prior to the addition of any stiffening. The critical load factors relative this base stress distribution cannot be reduced by the addition of any stiffening. If, however, as pointed out by Tarnai (1980) and illustrated in Example 5.2, the stiffening can alter the base stress distribution and increase the buckling loads without there being a corresponding increase in the buckling resistance.

6.8   Series of Independent Structural Systems

Theorem 7 (a)

Consider a series of independent structural systems $\left(K_i - \lambda_i K_{gi}\right)D = 0\Big|_{i=1}^{n}$ having minimum positive critical load factors $\lambda_{1\,min}^{+}, \lambda_{2\,min}^{+}, \ldots \ldots \lambda_{n\,min}^{+}$. All K_{gi} are taken as positive definite or indefinite. It follows that the composite structure, having stiffness matrix $K = \sum_i K_i$ and loading given by $K_g = \lambda \sum_i K_{gi}$, will have a minimum positive critical load factor given by $\lambda_{min}^{+} \geq \left(\lambda_{min}^{+}\right)_{min}$ where $\left(\lambda_{min}^{+}\right)_{min}$ is the minimum of $\lambda_{1\,min}^{+}, \lambda_{2\,min}^{+}, \ldots \ldots \lambda_{n\,min}^{+}$.

By inverting the problem as in Theorem 3 it can be established that $\lambda_{max}^{+} \leq \left(\lambda_{max}^{+}\right)_{max}$. However, maximum critical load are seldom of little practical interest.

Theorem 7 (b)

It is also possible to reverse this statement and to consider a series of independent structural systems $\left(K_i - \lambda_i K_{gi}\right)D = 0\Big|_{i=1}^{n}$ having maximum negative critical load factors $\lambda_{1\,max}^{-}, \lambda_{2\,max}^{-}, \ldots \ldots \lambda_{n\,max}^{-}$. All K_{gi} are taken as positive definite or indefinite. It follows that the composite structure, having stiffness matrix $K = \sum_i K_i$ and loading given by $K_g = \lambda \sum_i K_{gi}$, will have a maximum negative critical load factor given by $\lambda_{max}^{-} \geq \left(\lambda_{max}^{-}\right)_{max}$ where $\left(\lambda_{max}^{-}\right)_{max}$ is the maximum of $\lambda_{1\,max}^{-}, \lambda_{2\,max}^{-}, \ldots \ldots \lambda_{n\,max}^{-}$.

Theorems 7(a) and 7(b) are illustrated graphically in Fig 6.5.

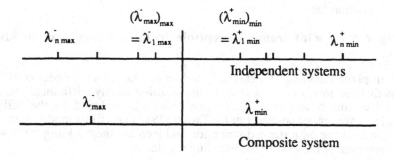

Fig 6.5   Minimum positive and maximum negative critical load factor of composite systems compared with minimum positive and maximum negative critical load factors of independent systems.

Proof (Theorem 7(a) only)

Consider the Rayleigh Quotient, ρ, for the composite structure in the situation where the eigen-vector, e, corresponding to λ^+_{min} has been chosen so that

$$\rho(e) = \frac{e^T K e}{e^T K_g e} = \lambda^+_{min}$$

For the independent structures, suppose that $e_1, e_2, \ldots\ldots e_n$ are the respective eigen-vectors corresponding to the minimum positive eigen-values given by

$$\lambda^+_{i\,min} = \frac{e_i^T K e_i}{e_i^T K_g e_i}\Bigg|^n_{i=1}$$

in which all numerators and denominators are positive. The numerators are positive because all K are positive definite and all denominators are positive because only positive eigen-values are under consideration.

Consider the individual structures buckling in the mode given by the vector e, ie, the vector which leads to the minimum positive eigen-value of the composite structure. All numerator terms, $e^T K_i e$ are ≥ 0, since all K_i are positive definite, but the denominator terms, $e^T K_{gi} e$, may be either positive or negative. Let j denote values of i for which $e^T K_{gi} e$ is positive and k denote values of i for which $e^T K_{gi} e$ is negative. Since $\lambda^+_{1\,min}, \lambda^+_{2\,min}, \ldots\ldots \lambda^+_{n\,min}$ are the minimum positive eigen-values for the individual systems then, for any j, $\rho_j(e) \geq \lambda^+_{j\,min}$.

It follows that

$$\sum_j \left[e^T K_j e\right] + \sum_k \left[e^T K_k e\right] = \lambda^+_{min} \sum_j \left[e^T K_{gj} e\right] - \lambda^+_{min} \sum_k \left[-e^T K_{gk} e\right]$$

$$= \sum_j \left[\frac{\lambda^+_{min}}{\rho_j(e)}\right] \left[e^T K_j e\right] - \sum_k \left[\frac{\lambda^+_{min}}{-\rho_k(e)}\right] \left[e^T K_k e\right]$$

where

$$\rho_j(e) = \frac{e^T K_j e}{e^T K_{gj} e}$$

By rearrangement

$$\frac{\lambda^+_{min}}{\left(\lambda^+_{imin}\right)_{min}} = \frac{\sum_j \left[e^T K_j e\right] + \sum_k \left[e^T K_k e\right]}{\left[\sum_j \frac{\left[\left(\lambda^+_{imin}\right)_{min}\right]}{\left[\rho_j(e)\right]}\left[e^T K_j e\right] - \sum_k \frac{\left[\left(\lambda^+_{imin}\right)_{min}\right]}{\left[-\rho_k(e)\right]}\left[e^T K_k e\right]\right]}$$

Since all bracketed terms are positive and $\rho_j(e) \geq \lambda^+_{min} \geq \left(\lambda^+_{imin}\right)_{min}$ it follows that $\lambda^+_{min} \geq \left(\lambda^+_{imin}\right)_{min}$. Note that $\frac{\left[\left(\lambda^+_{imin}\right)_{min}\right]}{\left[\rho_j(e)\right]} \leq 1$ and hence the denominator is less than the numerator.

6.9   Basis of Dunkerley's Equation

If, under load systems $\left.K_{gi}\right|^n_{i=1}$ a structure has minimum positive critical load factors $\left.\lambda^+_{imin}\right|^n_{i=1}$ then, under the combined load system $K_g = \sum\limits_{i=1}^n K_g$, the minimum positive critical load factor λ^+_{min} is not less than λ where $1/\lambda = \sum\limits_i 1/\lambda^+_{imin}$, ie, $1/\lambda^+_{min} \leq \sum\limits_i 1/\lambda^+_{imin}$.

Proof

Suppose that the structure has stiffness matrix K and consider a series of modified structures with stiffness matrices $\left.\left(\lambda/\lambda^+_{min}\right)K\right|^n_{i=1}$. If the modified structures are subjected, respectively, to load systems $\left.K_{gi}\right|^n_{i=1}$ then the least critical load factor for each structure will be λ. Let λ be chosen to have the value given by

$$\lambda = \sum_{i=1}^n \frac{1}{\lambda^+_{imin}}$$

so that a structure formed by the modified structures will have a stiffness matrix identical with the original structure. It follows from the previous result that the minimum positive critical load factor λ^+_{min} under the combined loading system satisfies the condition $\lambda^+_{min} \geq \lambda$.

6.9.1 Application of Dunkerley's Equation to Plate Buckling

Dunkerley's equation is a conservative version of the relationship, $1/\lambda^+_{min} \leq \sum\limits_i 1/\lambda^+_{imin}$. Noting that the expression may be rewritten as $1 \leq \sum\limits_i \lambda^+_{min}/\lambda^+_{imin}$ it follows that it is conservative to write $\sum\limits_i^n \lambda^+_{min}/\lambda^+_{imin} = 1$. Consider a flat plate subject simultaneously to critical direct stresses σ_{xcr}, σ_{ycr}, in-plane bending stress, σ_{bcr}, and shear stress, σ_{xycr}. Let the critical stresses when these stress components act alone be σ'_{xcr}, σ'_{ycr}, σ'_{bcr}, σ'_{xycr}. The Dunkerley formula takes the more familiar form

$$\frac{\sigma_{xcr}}{\sigma'_{xcr}} + \frac{\sigma_{ycr}}{\sigma'_{ycr}} + \frac{\sigma_{bcr}}{\sigma'_{bcr}} + \frac{\sigma_{xycr}}{\sigma'_{xycr}} = 1$$

A number of design codes use formulas of the Dunkerley type which always gives a conservative estimate of the critical load in situations where a complex

7 References

Bleich, F. (1952). Buckling Strength of Metal Structures, McGraw-Hill, New York.

Chen, WF and Lui, EM. (1987). Structural Stability - Theory and Implementation, Elsevier, New York.

Croll, JG and Walker, AC. (1972). Elements of Structural Stability, McMillan, London.

Gallagher, RH. (1976). Finite Element Analysis, Prentice-Hall, Englewood Cliffs.

Koiter, WT. (1967). On the Stability of Elastic Equilibrium, Dessertation, Delft, Holland, 1945. English Translation NASA, Tech Trans., F 10, 833.

Koiter, WT. (1976). Current Trends in the Theory of Buckling, Proc IUTAM Symposium Cambridge/USA 1974, Springer-Verlag, Heidelberg.

Milner, HR and Horne, MR. (1978) Elastic Critical Loads of Structural Systems, Int J Mech Sci, Vo 20, Permagon Press, Great Britain.

Murray, NW. (1984). Introduction to the theory of thin-walled structures, Clarendon Press, Oxford.

Poincare', H. (1885). Sur l'Euqilibre d'Une Masse Fluide Animee d'Un Mouvement de Rotation, Acta. math., 7, 259.

Renton, JD. (1967). Buckling of frames composed of thin-walled structures, Proc Conf Thin Walled Structures (Ed Chilver, AH), Chatto and Windus, London.

Tarnai, T (1980) Destabilizing effect of additional restraint on elastic bar structures. Int. J. mech. Sci., Permagon Press, Vol 22, pp379-390.

Thompson, JMT and Hunt, GW, (1973) A General Theory of Elastic Stability, Wiley, London.

Timoshenko, SP and Gere, JM (1961) Theory of Elastic Stability, Mc-Graw-Hill, New York.

Zaslavsky, A (1981) Discussion on destabilizing effect of additional restraint on elastic bar structures. Int. J. mech. Sci., Permagon Press, Vol 23, pp383-384.

APPENDIX A   STRAIN DEFINITIONS

A.1   Engineering Strain

Consider the filament shown in Fig A.1 of unstrained length, l_0, which moves and elongates to strained length, l_1. Engineering strain, Malvern (1973), is defined as:

$$\varepsilon_E = \frac{l_1 - l_0}{l_0} = \frac{change\ in\ length}{original\ length} \qquad\qquad A.1$$

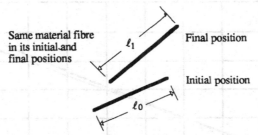

Same material fibre
in its initial and
final positions

ℓ_1

Final position

Initial position

ℓ_0

Fig A.1:   A filament before and after straining.

A.2   Green's Strain

Green's strain is based on the Pythagorean theorem and uses lengths squared. It is given by

$$\varepsilon_G = \frac{l_1^2 - l_0^2}{2 l_0^2} \qquad\qquad A.2$$

It is related to engineering strain by

$$\varepsilon_E = \frac{l_1 - l_0}{l_0} = \frac{(l_1 - l_0)(l_1 + l_0)}{l_0(l_1 + l_0)} = \frac{l_1^2 - l_0^2}{l_0^2(2 + \varepsilon_E)} = \varepsilon_G\left(\frac{1}{1 + \frac{1}{2}\varepsilon_E}\right) \qquad A.3$$

or $$\varepsilon_G = \varepsilon_E\left(1 + \tfrac{1}{2}\varepsilon_E\right) \qquad\qquad A.4$$

A.3   Comparison of Engineering and Green's Strain - Practical Significance

A question which arises concerns the effect of the choice of strain definition on the results of any analysis. The short answer is "very little".

Both Engineering and Green's strain are plotted in Fig A.2 against change in length per unit of original length. By observing the Fig A.2 plots and by substituting numerical values into eqn A.4, it is self-evidently true that, provided $\varepsilon_E \ll 1$, the practical difference is quite small. To gain some impression of the magnitude of strains which arise in practical structural analysis problems, consider typical structural grade steels (one of the more ductile of engineering materials) which yield at strains of less than 0.002 (corresponds to a stress level of approximately $400 MPa$). Suppose that a tension test is conducted on a structural grade steel to determine its Young's modulus by averaging the slope of the stress-strain curve between zero stress and the yield stress at $\varepsilon_E = 0.002$, $\varepsilon_G = 0.002 \times (1 + \frac{1}{2} \times 0.002)$ $= 0.002002$. At yield, the two strain measures differ only by 0.1%. Because of the low strain levels commonly encountered in most stability problems, debates concerning the strain definition are usually not significant.

In stability theory, both Engineering and Green's strains are used; Green's strain is preferred herein.

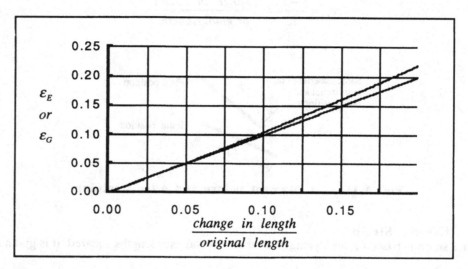

Fig A.2:   Comparison of Engineering and Green's strain to $\varepsilon_E = 0.2$.

A.4   Green's Strain Generalised and Defined in Terms of Displacements

The description below follows Fung (1973).

In Fig A.3 suppose that the material point P located by the position vector, a, in the undeformed state moves to point Q in the deformed state located by the position vector, x. The length of the undeformed filament, dl_0, is given by

$$dl_0^2 = da_1^2 + da_2^2 + da_3^2 = \delta_{ij} da_i da_j \qquad \text{A.5}$$

and the length of the deformed filament by

$$dl_1^2 = dx_1^2 + dx_2^2 + dx_3^2 = \delta_{ij} dx_i dx_j \qquad \text{A.6}$$

where δ_{ij} = Kronecker delta and repeated subscripts imply summation.

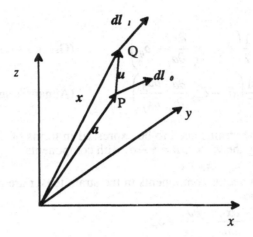

Fig A.3: Initial position, final position and displacement of a point.
P is the initial position and Q the final position.

Two systems of describing the positions of the same material point in a deformable body are in common use. In a Lagrangian system, the coordinates of the same material points in the deformed and undeformed configurations are given as functions of their position in the undeformed geometry. In an Eulerian system, the coordinates of the same material points in the deformed and undeformed configurations are given as functions of their positions in the deformed geometry.

If $x_i = x_i(a_1, a_2, a_3)$ and $a_i = a_i(x_1, x_2, x_3)$ for Langrangian and Eulerian descriptions, respectively, then it follows that $dx_i = \dfrac{\partial x_i}{\partial a_j} da_j$ and that $da_i = \dfrac{\partial a_i}{\partial x_j} dx_j$. On introduction into eqns A.5 and A.6, it follows that

$$dl_0^2 = \delta_{ij} da_i da_j = \delta_{ij} \frac{\partial a_i}{\partial x_l} \frac{\partial a_j}{\partial x_m} dx_l dx_m \tag{A.7}$$

$$dl_1^2 = \delta_{ij} dx_i dx_j = \delta_{ij} \frac{\partial x_i}{\partial a_l} \frac{\partial x_j}{\partial a_m} da_l da_m \tag{A.8}$$

The difference in the squares of the lengths of the line elements is finally given by:

Lagrangian description

$$dl_1^2 - dl_0^2 = \left(\delta_{\alpha\beta} \frac{\partial x_\alpha}{\partial a_i} \frac{\partial x_\beta}{\partial a_j} - \delta_{ij} \right) da_i da_j \tag{A.9}$$

Eulerian description

$$dl_1^2 - dl_0^2 = \left(\delta_{ij} - \delta_{\alpha\beta} \frac{\partial a_\alpha}{\partial x_i} \frac{\partial a_\beta}{\partial x_j} \right) dx_i dx_j \tag{A.10}$$

The strain tensors

$$E_{ij} = \frac{1}{2}\left(\delta_{\alpha\beta}\frac{\partial x_\alpha}{\partial a_i}\frac{\partial x_\beta}{\partial a_j} - \delta_{ij}\right) \qquad \text{(Green's strain tensor)} \qquad \text{A.11}$$

$$\varepsilon_{ij} = \frac{1}{2}\left(\delta_{ij} - \delta_{\alpha\beta}\frac{\partial a_\alpha}{\partial x_i}\frac{\partial a_\beta}{\partial x_j}\right) \qquad \text{(Almansi's strain tensor)} \qquad \text{A.12}$$

For structural analysis the strains need to be expressed in terms of displacements. Let the displacements be given by the vector, $u = x - a$, with components

$$u_\alpha = x_\alpha - a_\alpha \qquad\qquad \text{A.13}$$

The derivates of position vector components in the strain tensor are now given in terms of displacement derivatives as

$$\frac{\partial x_\alpha}{\partial a_i} = \frac{\partial u_\alpha}{\partial a_i} + \delta_{\alpha i} \qquad\qquad \text{A.14}$$

$$\frac{\partial a_\alpha}{\partial x_i} = \delta_{\alpha i} - \frac{\partial u_\alpha}{\partial x_i} \qquad\qquad \text{A.15}$$

From which the strain tensors are given by

$$E_{ij} = \frac{1}{2}\left[\delta_{\alpha\beta}\left(\frac{\partial u_\alpha}{\partial a_i} + \delta_{\alpha i}\right)\left(\frac{\partial u_\beta}{\partial a_j} + \delta_{\beta j}\right) - \delta_{ij}\right]$$

$$\qquad\qquad\qquad\qquad\qquad\qquad\qquad\qquad \text{A.16}$$

$$= \frac{1}{2}\left[\frac{\partial u_j}{\partial a_i} + \frac{\partial u_i}{\partial a_j} + \frac{\partial u_\alpha}{\partial a_i}\frac{\partial u_\alpha}{\partial a_j}\right]$$

$$\varepsilon_{ij} = \frac{1}{2}\left[\delta_{ij} - \delta_{\alpha\beta}\left(-\frac{\partial u_\alpha}{\partial x_i} + \delta_{\alpha i}\right)\left(-\frac{\partial u_\beta}{\partial x_j} + \delta_{\beta j}\right)\right]$$

$$\qquad\qquad\qquad\qquad\qquad\qquad\qquad\qquad \text{A.17}$$

$$= \frac{1}{2}\left[\frac{\partial u_j}{\partial x_i} + \frac{\partial u_i}{\partial x_j} - \frac{\partial u_\alpha}{\partial x_i}\frac{\partial u_\alpha}{\partial x_j}\right]$$

If unabridged notation is used [(x,y,z) replaces (x_1,x_2,x_3), (a,b,c) replaces (a_1,a_2,a_3), (u,v,w) replaces (u_1,u_2,u_3)] then the strain terms typically take the form:

$$E_{aa} = \frac{\partial u}{\partial a} + \frac{1}{2}\left[\left(\frac{\partial u}{\partial a}\right)^2 + \left(\frac{\partial v}{\partial a}\right)^2 + \left(\frac{\partial w}{\partial a}\right)^2\right]$$

$$\varepsilon_{xx} = \frac{\partial u}{\partial x} + \frac{1}{2}\left[\left(\frac{\partial u}{\partial x}\right)^2 + \left(\frac{\partial v}{\partial x}\right)^2 + \left(\frac{\partial w}{\partial x}\right)^2\right] \qquad \text{A.18}$$

$$E_{ab} = \frac{1}{2}\left[\frac{\partial u}{\partial a} + \frac{\partial v}{\partial b} + \left(\frac{\partial u}{\partial a}\frac{\partial u}{\partial a} + \frac{\partial v}{\partial a}\frac{\partial v}{\partial b} + \frac{\partial w}{\partial a}\frac{\partial w}{\partial b}\right)\right]$$

$$\varepsilon_{xy} = \frac{1}{2}\left[\frac{\partial u}{\partial y} + \frac{\partial v}{\partial x} - \left(\frac{\partial u}{\partial x}\frac{\partial u}{\partial y} + \frac{\partial v}{\partial x}\frac{\partial v}{\partial y} + \frac{\partial w}{\partial x}\frac{\partial w}{\partial y}\right)\right]$$

APPENDIX B   VIRTUAL WORK AND MINIMUM POTENTIAL ENERGY THEOREMS FOR BEAMS

B.1   Virtual Work Products and Theorems

Work Products are formed, typically, by multiplying force by displacement, force per unit length by displacement integrated over a length, stress by strain integrated over a volume, bending moment by curvature integrated over a length, etc. One of the product pair contains a force related term (force, force per unit length, stress, bending moment) and the other a displacement related term (displacement itself, strain, curvature) so that the product yields a quantity having work units (force x distance).

Virtual Work Products are formed with one of the work product pair (either the force or displacement related term) "virtual" and the other "real". Both pairs in the product belong to "fields", ie, their values are defined over the entire domain under analysis and selected on a defined basis. The "real" field is the force / stress or displacement / strain field of the actual solution to the problem or an approximation thereto. Therefore real force / stress fields must be statically admissible and real displacement / strain fields must be kinematically admissible. Virtual force / stress fields must be statically admissible but virtual displacement / strain fields need not be kinematically admissible although the strain field must be compatible with displacement field, ie, derivable from it.

In the application of work product methods to the solution of structural analysis problems strain energy and stress-strain (constitutive) relationships are not involved nor is the principle of conservation of energy. This independence of energy and constitutive laws is illustrated in Fig B.1.

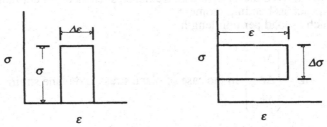

(a) Internal virtual work formed using actual stress and virtual strain.

(b) Internal virtual work formed using actual strain and virtual stress.

Fig B.1:   Definitions of virtual work products highlighting their independence of a stress-strain law.

Associate
d with the virtual work products are two virtual work theorems which differ with respect to which of the two terms in the work product is virtual. In the Virtual Displacement theorem it is the displacements and strains derived therefrom which are virtual and in the Virtual Force theorem it is the forces and stresses which are virtual.

B.2 Work Products formed from Virtual Displacements and Strains

With virtual work products formed using real forces / stresses and virtual displacements / strains the following apply.

i)   The "real" load and stress fields in the product meet statical admissibility requirements, in the case of a beam this means that the beam equation of equilibrium and the associated boundary conditions. As stated above, the "virtual" displacement and strain fields are not required to meet any condition other than requiring that the virtual strains be computed from the virtual displacements field by differentiation.

ii) Virtual work products of this type are formed by multiplying real internal stresses by virtual strains and real applied forces by virtual displacements to form, respectively, internal, V_I, and external, V_E, work products. The symbol Δ will be used to designate virtual displacements. The two products are given by

For a single component stress-strain field

$$V_I = \int_V (\Delta\varepsilon\sigma)dV \qquad\qquad\qquad\qquad\qquad \text{B.1a}$$

$$V_E = \int_S (\Delta v q)dS \qquad\qquad\qquad\qquad\qquad \text{B.1b}$$

For a multi component stress-strain field

$$V_I = \int_V (\Delta\varepsilon^T\sigma)dV \qquad\qquad\qquad\qquad\qquad \text{B.1c}$$

$$V_E = \int_S (\Delta u^T q)dS \qquad\qquad\qquad\qquad\qquad \text{B.1d}$$

where
Δv = virtual displacements, such as lateral beam displacements
$\Delta\varepsilon$ = virtual strains derived the virtual displacements, such as beam curvatures
σ = real stresses, such as bending moment
q = real loads, such as load per unit length

$$\sigma = \begin{Bmatrix} \sigma_{xx} \\ \sigma_{yy} \\ \sigma_{xy} \end{Bmatrix} \qquad \Delta\varepsilon = \begin{Bmatrix} \Delta\varepsilon_{xx} \\ \Delta\varepsilon_{yy} \\ \Delta\varepsilon_{xy} \end{Bmatrix} \text{ in the case of plane stress and plane strain}$$

iii) In the case of a beam in bending eqn B.1a needs to be interpreted as $V_I = \int_V (\Delta\varepsilon\sigma)dV$

$$= \int_L (\Delta\phi M_i)dL$$
where
$\sigma \equiv M_i$ = generalised stress
$\Delta\varepsilon \equiv \Delta\phi$ = generalised strain
$dV \equiv dL$ = indication of integration over length rather than volume

B.3   Work Products formed using Virtual Forces and Stresses

With virtual work products formed using real displacements / strains and virtual forces / stresses the following apply.

i)   The "real" displacement and strain fields in the product meet kinematical admissibility requirements. As stated above, the "virtual" force and stress fields are required to meet statical admissibility requirements, in the case of beams, eqn A1.

ii)   Virtual work products of this type are formed by multiplying real internal strains by virtual stresses and real displacements by virtual forces to form, respectively, internal, V_I, and external, V_E, work products.

For a single component stress-strain field

$$V_I = \int_V (\Delta \sigma \varepsilon) dV \qquad\qquad\qquad\qquad\qquad\qquad \text{B.2a}$$

$$V_E = \int_S (\Delta q) v dS \qquad\qquad\qquad\qquad\qquad\qquad \text{B.2b}$$

For a multi component stress-strain field

$$V_I = \int_V \left(\Delta \sigma^T \varepsilon\right) dV \qquad\qquad\qquad\qquad\qquad\qquad \text{B.2c}$$

$$V_E = \int_S \left(\Delta q^T u\right) dS \qquad\qquad\qquad\qquad\qquad\qquad \text{B.2d}$$

Virtual work products formed in this manner are commonly evaluated with a view to computing displacements from strains, eg, curvatures (generalised strains) from bending moments (generalised stresses) in a beam. The real strains (curvatures) can be either computed ($= M_i/EI$ if the bending moment distribution is known) or measured using a curvature-meter. The virtual force / stress field needs to be statically admissible relative to the virtual force field to arrive at a useful theorem.

iii)   In the case of a beam in bending eqn B.3a needs to be interpreted as $V_I = \int_V (\varepsilon \Delta \sigma) dV$
$= \int_L (\phi \Delta M_i) dL$.

B.4   Theorem of Virtual Displacements

B.4.1   Statement

For any real, statically admissible, stress field the external virtual work product arising from an arbitrary virtual displacement is equal to the internal virtual work product of the real stresses and the corresponding strains.

B.4.2   Single Beam

The proof is provided for the case of an isolated beam only.

Suppose that a statically admissible internal bending moment M_i has been determined by some means and that a work product is formed with a virtual displacement field Δv. The internal virtual work product is given using eqn B.1a.

$$V_I = \int_0^L M_i \Delta\phi dx \qquad\qquad \text{B.3}$$

where $\quad \Delta\phi = \dfrac{d^2\Delta v}{dx^2}$ is interpreted as the generalised virtual strain

$\qquad\quad M_i$ is interpreted as the generalised stress

Integrating once by parts leads to

$$\int_0^L M_i \Delta\phi dx = \int_0^L M_i \frac{d^2\Delta v}{dx^2} dx$$

$$= \int_0^L M_i d\left(\frac{d\Delta v}{dx}\right) = \left[M_i\left(\frac{d\Delta v}{dx}\right)\right]_0^L - \int_0^L \left(\frac{d\Delta v}{dx}\right)\left(\frac{dM_i}{dx}\right)dx \qquad \text{B.4}$$

$$= \left[M_i\Delta\theta\right]_0^L + \int_0^L \left(\frac{d\Delta v}{dx}\right)V_i dx$$

Integrating the second term by parts leads to

$$V_I = \left[M_i\Delta\theta + V_i\Delta v\right]_0^L - \int_0^L \left(\frac{dV_i}{dx}\right)\Delta v dx = \left[M_i\Delta\theta + V_i\Delta v\right]_0^L + \int_0^L q\Delta v dx$$

$$= M_{i(L)}\Delta\theta_2 - M_{i(0)}\Delta\theta_1 + V_{i(L)}\Delta v_2 - V_{i(0)}\Delta v_1 + \int_0^L q\Delta v dx \qquad \text{B.5a}$$

$$= M_1\Delta\theta_1 + M_2\Delta\theta_2 + V_1\Delta v_1 + V_2\Delta v_2 + \int_0^L q\Delta v dx = V_E$$

or

$$V_I = V_E \qquad\qquad \text{B.5b}$$

which establishes the proof which can be formally stated

$$\boxed{V_I = \int_0^L M_i \Delta\phi dx = M_1\Delta\theta_1 + M_2\Delta\theta_2 + V_1\Delta v_1 + V_2\Delta v2 + \int_0^L q\Delta v dx = V_E} \qquad \text{Proposition 1}$$

At the boundaries

$$M_1 = -(M_i)_{x=0} \quad M_2 = (M_i)_{x=L}$$
$$V_1 = -(V_i)_{x=0} \qquad V_2 = (V_i)_{x=L}$$

where positive curvature occurs when d^2v/dx^2 is positive and positive bending moment causes positive curvature; see Fig B.2a. In addition, positive shears are upwards on the right end. In Fig B.2b positive end forces and displacements are illustrated.

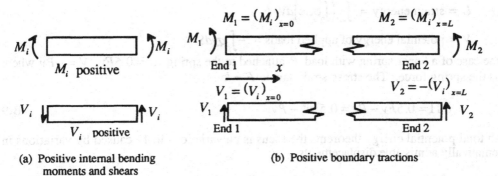

Fig B.2:   Sign convention.

B.5 Total Potential Energy

The theorem of stationary total potential energy imposes restrictions on the problem parameters in stress analysis that do not exist in the theorem of virtual displacements to which it is related. By definition, strain energy is computed from the area under the stress-strain diagram. It is therefore essential that stresses and strains be so related and that kinematic admissibility requirements be met. This was not required in the theorem of virtual displacements.

B.5.1   Potential Energy of the Applied Loads

As stated above it was unnecessary to introduce the concept of strain energy to arrive at work product principles. However, with energy principles, the notion of strain energy and potential energy of the applied loads is essential.

The concept is commonly introduced by considering a mass, m, in a gravitational field capable of generating an acceleration, g. The potential energy is $mgh = Fh$ where h is the height above some reference point. It is positive if the mass is raised against g. In structures the load application points on the boundary deflect under load and there is a loss of potential energy given by

$$V = -\sum Fv$$
B.6

If the applied loads are distributed over the surface then

$$V = -\int_S qv dS$$
B.7

and this encompasses point loads if these are assumed to be represented by Dirac distributions.

B.5.2   Total Potential Energy Expressions

The total potential energy of a system is given by

$$\Pi = U + V$$
B.8

where

$$U = \text{strain energy} = \int_{vol}\left(\int_0^\varepsilon \sigma d\varepsilon\right)dV$$

$$V = \text{potential energy of applied loads} = -\int_S qvdS$$

In the case of a linear spring with load P attached to the spring, $U = 0.5Fv$, $V = -Pv$ where F is the spring force. The stress-strain law is $F = kv$.

$$\Pi = 0.5Fv - Pv = 0.5kv^2 - Pv \qquad\qquad\qquad\qquad\qquad\qquad\qquad \text{B.9}$$

With total potential energy theorems the focus is on variations in Π caused by variations in kinematically admissible displacements.

In the case of a linearly elastic beam under distributed load

$$U = 0.5\int_0^L \phi M_i dx \qquad\qquad \text{and } V = -\int_0^L qvdS$$

where $M_i = \phi EI$

Hence

$$\Pi = 0.5\int_0^L EI\phi^2 dx - \int_0^L qvdS = 0.5\int_0^L \left(\frac{d^2v}{dx^2}\right)dx - \int_0^L qvdS \qquad \text{B.10}$$

B.5.3   Theorem of Minimum Total Potential Energy

For a stable elastic structure the total potential energy evaluated for all kinematically admissible displacements is minimum when the statical admissibility requirements are also met.

The term elastic does not imply that stress is proportional to strain but simply that a stress can be uniquely assigned for a given strain making it possible to compute the strain energy.

Proof

The total potential energy of the beam is given by

$$\Pi = 0.5\int_0^L EI\phi^2 dx - M_1\theta_1 - M_2\theta_2 - V_1v_1 - V_2v_2 - \int_0^L qvdx$$

$$\Pi + \delta\Pi = 0.5\int_0^L EI(\phi + \delta\phi)^2 dx$$

$$- M_1(\theta_1 + \delta\theta_1) - M_2(\theta_2 + \delta\theta_2) - V_1(v_1 + \delta v_1) - V_2(v_2 + \delta v_2)$$

$$- \int_0^L q(v + \delta v)dx$$

from which, by subtraction,

$$\delta\Pi = \int_0^L EI\phi\delta\phi dx - M_1\delta\theta_1 - M_2\delta\theta_2 - V_1\delta v_1 - V_2\delta v_2 - \int_0^L q\delta v dx = 0 \quad \text{B.11}$$

in view of the theorem of virtual displacements.

Now consider a displacement variation from an equilibrium position, Δv, leading to variations $\Delta \phi$ and $\Delta \Pi$ related by

$$\delta(\delta \Pi) = \delta^2 \Pi = \int_0^L EI(\delta \phi)^2 \, dx \qquad \qquad \text{B.12}$$

This is positive and establishes the fact the total potential energy is minimum when the member is in equilibrium.

THE RANKINE-MERCHANT LOAD AND ITS APPLICATION

M.R. Horne
University of Manchester, Manchester, UK

Relationships between the elastic critical loads, non-linear elastic load-deflection behaviour, rigid-plastic behaviour and elastic-plastic failure loads of frame structures are discussed. The concept of deteriorated critical loads is explained, and its value in obtaining an understanding of the criteria for failure of structures in the elastic-plastic range demonstrated by reference to examples. The use in the Rankine-Merchant formula of the two idealised load factors for elastic buckling and rigid-plastic collapse to obtain an approximation to the true elastic failure load is explained, and it is shown what idealisations have to be made to arrive at a formal proof. This enables conclusions to be reached over the situations in which the Rankine-Merchant load tends to give either a high or a low estimate of the true failure load. Correlations between the Rankine-Merchant loads and the true collapse loads obtained in both theoretical and experimental investigations are presented. Finally proposals are made to enable the Rankine-Merchant load, either in its original or in a modified form, to be used reliably for design purposes.

3.1 THE RIGID-PLASTIC AND ELASTIC-PERFECTLY PLASTIC IDEALISATIONS OF STRUCTURAL BEHAVIOUR

3.1.1 Introduction

We will be concerned with the derivation and use of the Rankine-Merchant load as an approximation to the elastic-plastic failure loads of frame structures, ie, frames which depend on their resistance to loads primarily on the flexural bending resistance of their members. While computer programs are available which are capable of carrying out full analyses of quite extensive frames into the elastic-plastic range and up to collapse, there is an advantage in exploring approximate procedures which depend on analytical concepts based on simplified models of structural behaviour. Two advantages can follow from this approach. First, the simplified models can frequently be solved by hand procedures, thus enabling the engineer to avoid the danger of merely accepting the results of computer programs which can be prone to major errors. Errors tend to arise, not so much because of faults in the programs themselves, but more frequently because of the readiness with which mistakes may be made by the user who is tempted merely to use a program as a "black box" without fully understanding the basic assumptions involved. Secondly, the way in which the simplified failure load solutions interact to encompass an approximation to the total behaviour can contribute to a much better understanding of the structural principles involved.

Most of the material in this course is concerned with theoretical concepts arising from the application of an assumed linear, reversible elastic relationship between stress and strain. We must first consider the various ways in which it is possible to introduce plastic, irreversible strain.

3.1.2 The physical approximations of plastic theory

Interest in plastic theory from the point of view of structural practice has arisen because of the particular stress-strain characteristics of structural mild steel. In the range of strains which are of structural importance, the stress-strain relation may be represented as in Fig 1(a). Elastic behaviour OA up to an upper evanescent yield stress f_U is followed by pure plastic strain at the lower yield stress f_L, the range of pure plastic strain BC being, for *mild* structural steel, some 10 to 20 times the elastic strain at yield.

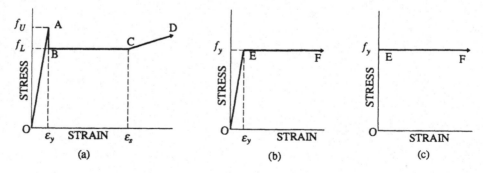

Figure 1   a)   **Stress-strain relation for mild steel**
                b)   **Idealised elastic-plastic stress-strain relation**
                c)   **Rigid-plastic idealization**

At C, "strain-hardening" sets in, and there is a rise of stress, up to an ultimate material failure stress f_{ULT} at a rate which is only a few percent of the rate of increase of stress in the elastic range OA. The upper yield stress is, in practice, for rolled or fabricated structural sections, entirely masked by the presence of residual stress, simply because of the sequential way in which nearly adjacent longitudinal fibres reach yield point. Since also the strain ε_s at the beginning of strain-hardening is large, the elastic-pure plastic idealization of the stress-strain relation, Fig 1(b), leads to descriptions of structural behaviour which are sufficiently accurate for most practical purposes. Applied to the bending of a structural section, the stress-strain relation of Fig 1(b) leads to a theoretical *moment-curvature* relation which closely mirrors the stress-strain relation, except that the corner at E becomes rounded because of the gradual spread of plastic deformation from the extreme fibre towards the neutral axis. This rounding can however be ignored, and so we come to the concept of a *rigid plastic hinge* which can undergo indefinite rotation at a constant *plastic moment of resistance* M_p.

An assumed structural behaviour in which a member remains elastic except where a plastic hinge forms provides the basis for the great majority of elastic-plastic structural computer programs. To avoid the complexities of such analyses, the *plastic theory of structures* is based on the further idealization of the *rigid-plastic* stress-strain relation shown in Fig 1(c). In frame structures which resist external loads primarily by bending action, deformation is than entirely confined to plastic hinges, the intermediate portions of members remaining rigid. No deformation of the structure can therefore take place until sufficient plastic hinges have formed to transform the structure, or part of it, into a mechanism.

3.1.3   Fundamental principles of rigid-plastic theory

Fundamental theorems relating to the estimation of rigid-plastic failure loads for plane rigid frames have been established, namely the *uniqueness theorem* and the *minimum and maximum principles* (Horne[1], Greenberg and Prager[2]). Many techniques for estimating rigid-plastic collapse loads, based on these fundamental theorems, are available, for example, see [3], [4], [5]. It is imagined that a rigid-plastic structure is subjected to a series of loads which, while varying in magnitude, bear a constant ratio each to each (monotonic loading). The general level may than be characterised by a single load factor, which will be denoted by λ.

According to the *uniqueness theorem*, if at any load factor λ, a bending moment distribution can be found satisfying the three conditions of (i) equilibrium, (ii) mechanism and (iii) yield, then λ is the rigid-plastic collapse load factor λ_p. The equilibrium condition is that the postulated internal moments and accompanying forces shall be in equilibrium with the external loads. According to the mechanism condition, there shall be sufficient plastic hinges to transform the structure, or part of it, into a mechanism and, finally, according to the yield condition, the full plastic moment shall nowhere be exceeded.

The evaluation of collapse loads is aided by the use of the *minimum* and *maximum* principles. Using the *minimum principle*, the collapse load factor is obtained as the least value derived by equating the internal and external work (which ensures that the structure is in equilibrium) for any postulated collapse mechanism. Using the *maximum principle* the collapse load factor is the greatest factor for which it is possible to postulate a state of internal moments and forces which satisfies both the yield and equilibrium conditions.

3.1.4   The effect of finite deflections on the rigid-plastic collapse load

The rigid-plastic collapse load is usually calculated for the state of vanishingly small deformations of the collapse mechanism. For the single-storey portal frame in Fig 2(a), loaded as shown and having a uniform plastic moment $M_p = WL$, the collapse mechanism is as shown in Fig 2(b) (see [6], chapter 4). For infinitely small displacements, the work equation becomes

$$\lambda_p W(L\theta) + 2\lambda_p W\left(\frac{L}{3}\theta\right) + 2\lambda_p W\left(\frac{L}{3}\frac{\theta}{2}\right) = M_p\left(\frac{3\theta}{2} + \frac{3\theta}{2}\right) \tag{1}$$

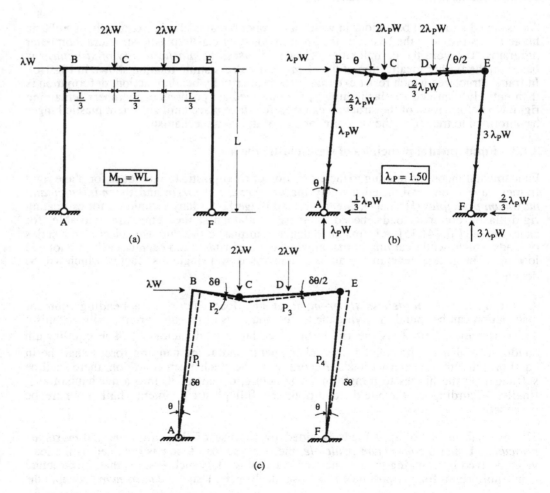

Figure 2 **Rigid-plastic collapse of single-bay, single story portal frame**

leading to a value for the collapse load factor of $\lambda_p = 1.50$. The angle θ in equation (1) represents both a virtual and the actual collapse mechanism displacement taken as small enough that it has no effect on the internal stress resultants (internal axial forces, bending moments and shears).

Consider now the value of the load factor after a finite deformation of the rigid plastic mechanism characterised by a finite value of the rotation θ now taken to be the real displacement pattern. The virtual work equation is formed with virtual deformation $\delta\theta$, Fig 2(c), assumed to be proportional to the real finite rotations θ. If the effects of the axial thrusts in the members on the external work terms are ignored, the work equation is simply (1) above with θ replaced by $\delta\theta$, giving $\lambda = \lambda_p = 1.50$. However, because of the presence of the axial load thrusts P_1, P_2, P_3, and P_4, see Fig 2(c), additional external work terms are introduced as follows,

$$P_1 L\theta\delta\theta + P_2 \frac{L}{3}\theta\delta\theta + P_3 \frac{2L}{3}\frac{\theta}{2}\frac{\delta\theta}{2} + P_4 L\theta\delta\theta$$

The exact values of the thrusts in the finite deformation state are tedious to calculate, but may be assumed to remain quite closely proportional to the load factor λ. Hence

$$P_1 = \lambda W, \; P_2 = P_3 = \tfrac{2}{3}\lambda W \text{ and } P_4 = 3\lambda W$$

The complete work equation thus becomes, after the addition of the further work terms,

$$\begin{aligned}
\Big[\lambda W L\delta\theta &+ 2\lambda W \frac{L}{3}\delta\theta + 2\lambda W \frac{L}{3}\frac{\delta\theta}{2} + \lambda W L\theta\delta\theta \\
&+ \frac{2}{3}\lambda W \frac{L\theta}{3}\delta\theta + \frac{2}{3}\lambda W \frac{L\theta}{6}\delta\theta + 3\lambda W L\theta\delta\theta \Big] = M_p \Big[\frac{3\delta\theta}{2} + \frac{3\delta\theta}{2} \Big]
\end{aligned} \qquad (2)$$

Substituting $M_p = WL$ and simplifying,

$$\lambda = \frac{1.5}{1 + \dfrac{13}{6}\theta} = \frac{1.5}{1 + \dfrac{13}{6}\dfrac{\Delta}{L}} \qquad (3)$$

where Δ is the sway deflection of the beam BCDE.

In more general terms, for any collapse mechanism, the effects of finite deformations may be included by using a complete work equation in the form

$$\lambda \sum W\delta\Delta + \lambda \sum PL\theta\delta\theta = \sum M_p \delta\phi \qquad (4)$$

Here, from the finite plastic collapse deformation state, loads deflect through additional displacements $\delta\Delta$, rigid links of length L in the collapse mechanism, carrying thrusts P, already rotated through angles θ, rotate through virtual angles $\delta\theta$ and plastic hinges, with plastic moments of resistance M_p, rotate through virtual angles $\delta\phi$. Assuming that the incremental displacements and rotations are proportional to the already achieved finite values, the work equation (4) may be replaced by

$$\lambda \sum W\Delta + \lambda \sum PL\theta^2 = \sum M_p\phi \tag{5}$$

When the effects of finite deformations are neglected, the collapse load factor λ_p is given by

$$\lambda_p \sum W\Delta = \sum M_p\phi \tag{6}$$

whence it follows that

$$\lambda = \frac{\lambda_p}{1 + \lambda_p\left[\sum(PL\theta^2)/\sum(M_p\phi)\right]} \tag{7}$$

The rigid-plastic relationship for the frame in Fig 2(a) as given by equation (3) is shown by the lower curve AC in Fig 3. Line AB is a plot of the relationship which applies when the effect of finite deformations on the equilibrium state is ignored. The upper curve AD represents the accurately calculated load-deflection relationship which allows for the change of member axial load with deformation and for accurately calculated changes of geometry. Up to quite large displacements, AC is indistinguishable from AD, and is as sufficiently accurate for practical purposes. Equation (7) is similarly sufficiently accurate for all practical purposes.

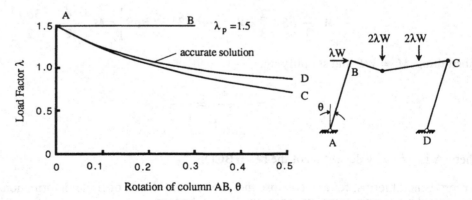

Figure 3   Rigid-plastic deflection curves for the frame in Figure 2

3.1.5 Elastic-plastic compared with rigid-plastic behaviour

A comparison between the above rigid-plastic type of behaviour with that derived by assuming idealised elastic-plastic behaviour, Fig 1(b), is illustrated in fig 5 for a similar portal frame, Fig 4, but one having fixed foundations and differently loaded, with the slenderness ratio L/r for in-plane bending of the assumed steel columns being of the order of 400. The frame is very much more slender than would be the case for any practical single-storey frame, but we are here concerned with clarifying the general principles of behaviour. The effects can become of potential practical importance for multi-storey frames, the principles remaining the same.

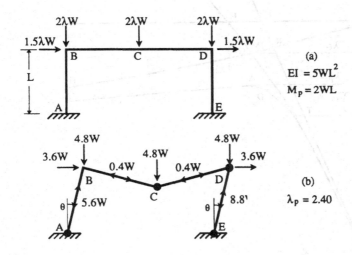

Figure 4   a) Loaded portal frame   b) Plastic collapse state

The drooping curve RGN in Fig 5 is the rigid-plastic behaviour of the frame collapsing in its collapse mechanism, Fig 4(b), to be compared with the non-drooping relationship RH when the effects of finite deformations are neglected. When a linear elastic analysis is carried out for the frame, again ignoring all the effects of change of geometry on the conditions of internal equilibrium, the straight line relationship OKL is obtained. When a complete non-linear elastic analysis is performed, allowing for all change of geometry effects including both the effects of axial forces on the stiffnesses of the members and the effects of the sway deflections on the overall equilibrium conditions for the frame, the curve OGH is obtained. When a complete elastic-plastic analysis is performed, allowing for all change of geometry effects, the load-deflection relationship becomes OEDAQ. Instead of plastic hinges forming simultaneously at A, C, D and E when the collapse load factor 2.40 is reached, as given by the rigid-plastic solution point R, the plastic hinges form sequentially in the order E, D, A and only finally at C, off the scale of the load-deflection curve beyond Q in Fig 5.

For the frame as indicated, the actual failure load at a load factor of 1.80 is thus 25% below the rigid-plastic failure load of 2.40. While it is possible, by computer program, to follow the step-by-step formation of the plastic hinges and thus to compute the failure load, it is of interest to find out whether there exist relationships between the other idealised load-deflection relationships which enable some approximate estimate to be made of the failure load without the necessity of carrying out such a procedure. It is also of interest to ascertain what theoretical considerations govern the point on the elastic-plastic load-deflection relationship, such as point A in this example, when the peak load is attained. These matters will be explored in succeeding sections.

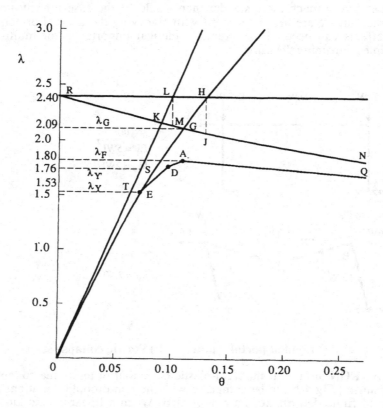

Figure 5   Load-deflection relationships for the frame in Figure 4

3.2 THE RIGID-PLASTIC AND ELASTIC-PERFECTLY PLASTIC IDEALIZATIONS OF STRUCTURAL BEHAVIOUR

3.2.1 "Compensated" and "uncompensated" loading

The distinctions between, on the one hand, the rigid-plastic and linear elastic load-deflection relations that neglect change of geometry effects, eg, RH and OL respectively in Fig 5, and the corresponding relations that allow for change of geometry, RGH and OGH respectively, need to be defined more rigorously by reference to the nature of the corresponding loadings applied to the frame in question.

We consider a structure, Fig 6(a), subjected to external loads acting at any points in any directions. The structure is imagined to be a 'frame structure', that is, a structure which is dependent on rigid as opposed to pinned joints to prevent it becoming a mechanism and is not triangulated. The loads λQ_j are divided into components λR_j perpendicular to the members on which they act and λS_j parallel to the members. The axial loads in the members due to the loads λQ_j will be denoted by λP_i. The values of P_i will not remain absolutely constant during the various stages of elastic and plastic deformation, since axial loads will not remain exactly proportional to external loads. The approximation will be made that such variations in P_i may be neglected, so that P_i denotes a constant set of axial forces, to which is applied the common load factor λ.

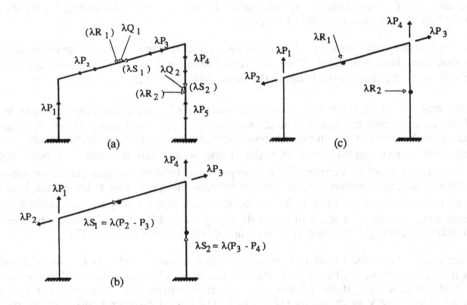

Figure 6 a) Load system λQ_j and induced axial loads
b) Compensating loads
c) Load system (a) + compensating loads (b).

The effects of change of geometry on the equations of equilibrium and, for elastic and elastic-plastic structures, on the flexural equations for the members are associated with the presence of axial loads λP_i in the members. It is desirable to be able to isolate such effects on the behaviour of the frame by postulating suitable external forces $\lambda T_k = -\lambda P_i$ applied at the joints and at the points in the structure where the loads λQ_j are applied, Fig 6(b). The values of these forces at the loading points $(j = k)$ are equal and opposite to the components λS_j, parallel to the appropriate member, of the respective applied loads λQ_j. Hence, within the lengths of members, the structure is loaded by only by forces λR_j which are perpendicular to the members and by forces at the joints which eliminate axial loads in all of the members, Fig 6(c). This type of loading will be referred to as "*compensated loading"*, since the axial loads in the members are reduced to zero with such additional loads acting. The original loading is conversely referred to as *"uncompensated loading"*. The loads λT_k are the *"compensating loads"*. Loads $-\lambda T_k$ would, if applied, create axial loads equal to those caused by uncompensated loading, without directly causing any flexural deformation apart from the secondary moments due to direct axial changes of lengths. For the type of frame under consideration, these secondary moments may be ignored and these negative compensating loads may be referred to as *"equivalent axial loading"*.

3.2.2   The idealized load-deflection relations

We are now in a position to consider all the possible "idealized" load-deflection relationships, derived from the application of the above types of loading, for rigid-plastic, elastic and elastic-plastic modes of response.

Fig 7 illustrates the various types of behaviour that ensue, and the general pattern of relationships between them. The load factor λ is plotted against some characteristic deflection Δ, the most suitable being the sway deflection at the top of the frame.

Under equivalent axial loading, involving forces $-\lambda T_k$, a rigid-plastic structure will not deform at all within the range of loads relevant to the present study. If one increased the loading to the point at which one or more members reached their "squash loads" (= cross-section area times yield stress σ_y), the frame would fail in some, for our purposes, irrelevant catastrophic manner. Under compensated loading, a rigid-plastic structure will collapse in a hinge mechanism, the load factor remaining constant at the plastic load factor λ_p as Δ increases, LN in Fig 7. As has been seen, under uncompensated loading, a rigid-plastic structure deflects according to a drooping curve LK, the relationship between load and deflection being expressed by an equation of the form of equation (7).

Under equivalent axial loading, a structure which remains perfectly elastic will show no flexural deformation until its lowest elastic critical buckling load factor λ (eigenvalue) is attained. At this load, if the effects of *gross* deformations are neglected, the deflections increase indefinitely as shown by HJ in Fig 7. The lateral displacements v of the members are those of the first elastic critical mode $v = v_1$, where the v are measured in the directions perpendicular to the longitudinal axis of any member. There exists a series of higher critical load factors (or eigenvalues) $\lambda_1 \leq \lambda_2 \leq \lambda_3 \leq \lambda_4 \leq ....$ and corresponding critical modes v_1, v_2, v_3, v_4 etc into which the structure could theoretically deform if the lower critical modes were suppressed.

Under compensated loading, with lateral loads applied but all axial forces in the members eliminated, the load-deflection relationship for a purely elastic structure becomes linear according to the straight line OC in Fig 7. The deflections, v, may be expressed as a series in terms of the critical eigen-modes v_1, v_2, v_3 etc, namely,

$$v = \lambda \left(a_1 v_1 + a_2 v_2 + a_3 v_3 + ... \right) \tag{8}$$

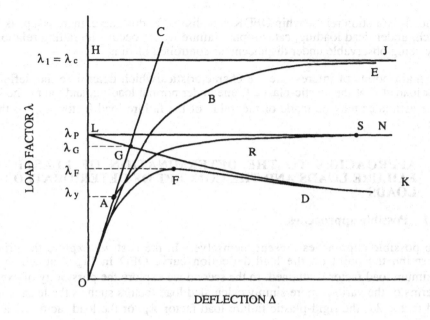

Figure 7   Load-deflection relationships for various idealized states
         LN   rigid-plastic (compensated loading)
         LK   rigid-plastic (uncompensated loading)
         OC   linear elastic response (ie compensated loading)
         OBE   elastic response (uncompensated, ie, normal loading)
         ORS   elastic-plastic response (compensated loading)
         OFD   elastic-plastic response (uncompensated, ie, actual response).

The curve OBE in Fig 7 is the load-deflection curve of a perfectly elastic frame subjected to uncompensated loading. This curve is tangential to the linear elastic line OC at O and asymptotic to HJ at the lowest critical load factor λ_1. It may be shown that the displacements v according to this curve can be expressed in terms of the eigen-modes of equation (8) as

$$v = \frac{a_1}{1 - \lambda/\lambda_1} v_1 + \frac{a_2}{1 - \lambda/\lambda_2} v_2 + \ldots\ldots\ldots \tag{9}$$

Finally, we consider a structure in which the material is assumed to have the more realistic elastic-plastic relationship of Fig 1(b). Under compensated loading, the behaviour follows the curve ORS, initially tangential to OC but rising towards the simple rigid-plastic collapse behaviour LN with which it merges as soon as the required number of mechanism hinges have formed at S. Under the normal, uncompensated loading, ie, the actual loading, the load-deflection relationship OFDK is realised. On this curve, there is a peak load at F at which, under dead loading, catastrophic failure would occur, the falling relationship FDK only being observable under displacement-controlled loading.

The main points of interest are the characteristics which determine the definition of the peak load at F of the elastic-plastic frame under normal loading, and the methods by which some estimates may be made of the value of the failure load factor λ_F. To this we now turn.

3.3 APPROACHES TO THE DETERMINATION OF ELASTIC-PLASTIC FAILURE LOADS AND THE CONCEPT OF DETERIORATED CRITICAL LOADS

3.3.1 Possible approaches

Two possible approaches present themselves. In the first we explore the criteria which determine the point on the load-deflection curve OFD in Fig 7 at which the actual maximum load factor is attained. In the second we explore the possibility of expressing λ_F in terms of the various more simply calculated load factors such as the least elastic critical load factor λ_C, the rigid-plastic failure load factor λ_P, or the load factor λ_y at which the yield stress σ_y is theoretically first reached in the perfectly elastic structure - point A in Fig 7. This latter approach is the one which led Merchant [8] to propose the formula now known as the Rankine-Merchant load, which will be considered in Section 3.4.1. We first consider the former approach.

3.3.2 Conditions for elastic-plastic failure

For a completely elastic structure, let $U = U_W + U_E$ where U is the total energy of the system, U_W is the potential energy of the applied load system and U_E is the strain energy of the structure. Consider a typical point on the load-deflection curve OCB of such a structure, Fig 8, where μ is some convenient deflection parameter such as the sway deflection at the top of a multi-storey frame. The variation of U with any change in the deformation state is itself stationary, ie,

$$\frac{dU}{d\mu} = 0 \tag{10}$$

Since the equilibrium is also stable for a point such as A on the rising part of the curve in Fig 8 up to but not including C, positive work will have to be done by some additional external agency to cause a displacement of the structure away from the equilibrium state. For such a stable structure therefore, the total energy of the system is at a minimum, Fig 9(a), and so

$$\frac{d^2U}{d\mu^2} > 0 \tag{11}$$

On the falling part of the load-deflection curve, such as CB in Fig 8, the structure is unstable and $\frac{d^2U}{d\mu^2} < 0$, Fig 9(c), while, at the peak load point C, Fig 8, the structure is in neutral equilibrium and $\frac{d^2U}{d\mu^2} = 0$, Fig 9(b). This last criterion is the same as that governing the attainment of the first elastic critical factor λ_1.

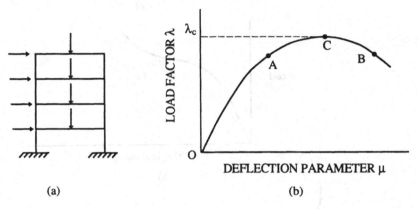

(a)   (b)

Figure 8   Elastic load-deflection curve for a frame structure.

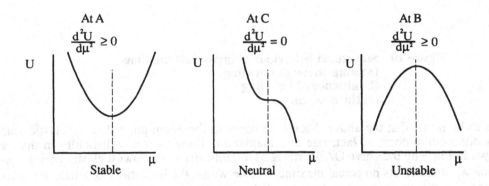

Figure 9 Stability criteria for elastic structures

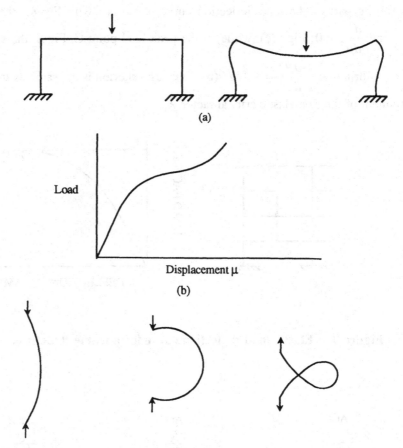

Figure 10   Structural behaviour at gross deformations
(a) snap-through buckling
(b) absence of buckling
(c) the elastica.

It is to be noted that the above discussion contains the assumption that the elastic load-deflection curve does, in fact, rise to a maximum. There is here a difficulty in that, as shown in Fig 7 by the curve OABE rising asymptotically to the lowest elastic critical load factor λ_1, there exists no actual maximum value within the limitations of small deflection theory. The discrepancy occurs because of the complications that arise with the treatment of *gross* deflections. Depending on the form of the structure, in the *gross* deflection regime, the elastic curve may either continue to an actual peak, as in so-called "snap-through" structures such as in that shown in Fig 10(a) or may, in fact, begin to show a load-deflection curve that begins to rise at an increased rate as is the case of the "elastica" deformation states, Fig 10(b), of a buckling strut, Fig 10(c). The type of case where a peak load is attained is used here since it provides a convenient introduction to the discussion of the failure criteria for elastic-plastic structures.

Consider now the load-deflection curve OAFD, Fig 7, for an elastic-plastic structure. We include in the total energy of the system not only the potential energy of the applied loads U_W and the elastic strain energy U_E of the deformed structure, but also the total energy U_P absorbed up to that point in the loading-deformation history of the loaded structural system. The energy absorbed in plastic deformation depends on the loading path for the structure, so that the system is no longer a conservative one. We consider a "total energy" quantity U_N including all these energy quantities, so that

$$U_N = U_W + U_E + U_P \tag{12}$$

For equilibrium of the structure the condition $\dfrac{dU_N}{d\mu} = 0$ applies at all points on the load-deflection curve OAFD in Fig 7. Moreover $\dfrac{d^2U_N}{d\mu^2} > 0$ before the failure load is reached at F, $\dfrac{d^2U_N}{d\mu^2} < 0$ on the falling part FD of the curve and $\dfrac{d^2U_N}{d\mu^2} = 0$ at the failure load at F when the load factor is λ_F. However, in the plastic zones of the structure, the stress remains constant at the yield value, and so necessarily $\dfrac{d^2U_P}{d\mu^2} = 0$. It follows form equation (12) that the condition at the failure load at F becomes

$$\frac{d^2(U_W + U_E)}{d\mu^2} = 0 \tag{13}$$

Since the elastic strain energy in the plastic zones of the structure is constant, it follows that the failure criterion of the elastic-plastic structure is identical, as far as the second differentials are concerned, with that for the same structure with those parts of the structure that are deforming plastically eliminated. The structure in this depleted condition is termed the *deteriorated structure*, and the corresponding elastic critical load is called the *deteriorated critical load*. If, for example, a plastic hinge has formed in an elastic-plastic structure, the hinge provides zero increase of resistance moment for an increase of rotation and can be regarded as a pinned joint when computing the deteriorated critical load. As the load on a structure increases and more and more plastic hinges are formed, the *deteriorated critical load factor* decreases continuously until it is depressed down to or below the load factor corresponding to the actual load applied. It is at this stage that the structure becomes unstable and the limiting load factor λ_F determined.

3.3.3   The use of the deteriorated elastic critical load criterion

The development of the deteriorated critical load concept to explain the behaviour of elastic-plastic frame structures was due to Wood [9]. As has already been explained, the behaviour of frames which resist external loads primarily by bending action can be followed with sufficient accuracy by assuming that plasticity is confined to plastic hinges, thus ignoring the existence of plastic zones in the outer parts of these sections of the members which are not fully plastic. For such a frame, the behaviour can be envisaged by reference to the load-deflection relationships shown in Fig 11. OAGH is the curve calculated on the assumption of indefinite elastic behaviour, and rises asymptotically to the line $\lambda = \lambda_c = \lambda_1$ where λ_1 is the lowest elastic critical load factor. The first plastic hinge is formed at A, defining the "first yield" load factor, λ_y. Thereafter, until the next hinge forms at B, the load-deflection curve follows the curve AJK. This rises asymptotically towards the load factor level λ_{D1}, the first deteriorated critical load level calculated for the original elastic structure but with a pinned hinge inserted where the first plastic hinge has formed in the elastic-plastic structure. After B on the load-deflection curve, the relationship rises towards the critical load factor line λ_{D2} corresponding to the deteriorated structure with two hinges. In the case illustrated, the formation of the third hinge at point C on the load-deflection curve causes the deteriorated critical load factor to fall below the current value, and the theoretical load-deflection curve now approaches $\lambda = \lambda_{D3}$ from above. If a complete plastic mechanism occurs at the formation of the fourth hinge at D, then this is a point on the rigid-plastic mechanism line FE. The failure load factor λ_F thus occurs at point C where the deteriorated critical load changes from being *above* the current load factor to falling *below*.

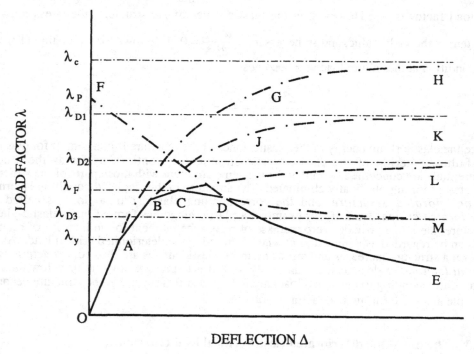

Figure 11   Load-deflection relationship illustrating deteriorated critical loads.

3.3.4   Example of a deteriorated critical load analysis

We consider the two-storey, single-bay frame illustrated in Fig 12(a). The frame is composed throughout of a Universal Column Section 152 x 152 UC 30, bent about its minor axis. Section and material properties follow

second moment of area	$558 cm^4$
yield stress	$240 N/mm^2$
plastic modulus, minor axis	$111.2 cm^3$
full plastic moment	$26.69 kNm$
elastic modulus	$206 kN/mm^2$

Under the "working loads" ($\lambda = 1.000$), shown in Fig 12(a) and expressed in kN, the deflected form is as indicated, the deflections being shown in metres. It will be seen that the frame is a slender one, these deflections at working load level being much higher than would be tolerated in practice. A slender frame has been adopted in order clearly to illustrate the principles involved.

Rigid-plastic collapse occurs at a load factor of $\lambda_p = 1.996$, the collapse mode being as shown in Fig 12(b). It will be noted that there are 6 hinges in the collapse mode.

The relevant elastic critical load is the load at which failure would occur under axial loading only, achieved by the division of all the vertical loads between the two columns, the effects of any axial loads in the beams being entirely negligible. This elastic critical load $\lambda_c = \lambda_1$ is found to have a value 6.727, the elastic critical deflection mode being as shown in Fig 12(c). It will be noted that the ratio of sway deflection in the upper storey Δ_2 to that in the lower storey Δ_1 differs between these three deflection states, being 0.88, 1.00 and 0.67 respectively.

The load-deflection relationships for the three assumptions of linear elastic response (compensated loading), non-linear elastic response (uncompensated loading) and elastic-plastic response (uncompensated loading) are shown in Fig 13. The latter relationship shows the stages at which the successive hinges are formed at the respective positions shown in Fig 14(a). It will be seen how an abrupt change of slope of the elastic-plastic response curve occurs with each formation of a new hinge, reflecting the step-by-step deterioration of the critical load. The maximum load occurs on the formation of the third hinge, ie, long before a complete plastic hinge mechanism has formed, the value of the peak load factor being 1.551, 22.3% below the rigid-plastic collapse load. The final hinge mechanism involves the lower storey only, as shown in Fig 14(b).

The concept of deteriorated critical loads is a very useful one in obtaining an understanding of how the progressive development of plasticity within a structure finally leads to collapse, and provides a criterion for the attainment of the peak load which can be used in the computer analysis, step-by-step to failure. This is the same as that used for identifying the attainment of a critical elastic load, namely that the stiffness matrix for the deteriorated elastic structure changes from being positive to being negative. However, as will be seen in Fig 13, the failure state can be reached when only a very few plastic hinges have formed and, since there is no means of finding out which hinges will have formed round about the point of maximum load, the deteriorated critical load criterion is not of itself of any great value in determining the failure load except within a step-by-step elastic-plastic analysis.

In the next section we will deal with the alternative approach in which the load parameters appearing in Fig 7 relevant to the various idealised forms of structural response are used to derive an approximation to the failure load factor λ_F.

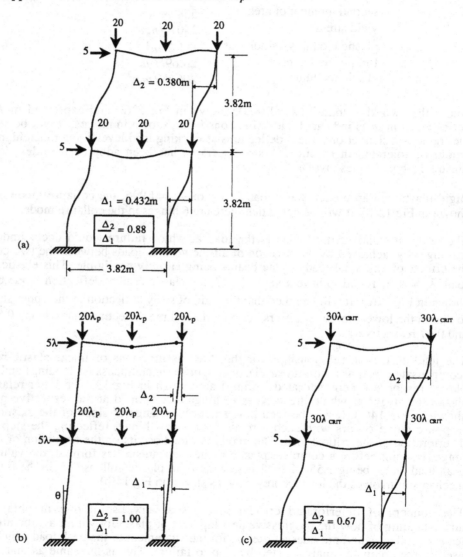

Figure 12 Loads and displacement modes for a two-storey frame
a) linear elastic analysis,
b) rigid-plastic collapse,
c) elastic critical mode.

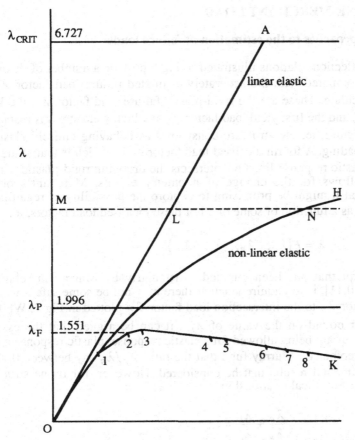

Figure 13 Load deflection relationships for frame in Figure 12 (deflection $(\Delta_1 + \Delta_2)$ at top of frame).

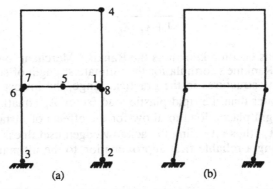

Figure 14 Formation of plastic hinges in the frame in Figure 12
a) Order of formation   b) Final active collapse mode.

3.4   THE RANKINE-MERCHANT LOAD

3.4.1   Empirical approaches to the estimation of failure loads

The idealised load-deflection relations illustrated in Fig 7 provide a number of theoretical load factors which, compared with the accurately estimated failure load factor λ_F, are relatively easy to calculate. These are the rigid-plastic failure load factor λ_P, the lowest critical load factor λ_c, and the first yield load factor λ_y at which yield stress is reached for the linear elastic structure, ie, the structure considered as behaving entirely elastically under *compensated* loading. A fourth idealised load factor is λ_G which is that at the point G where the linear elastic response line OC intersects the drooping rigid-plastic response curve LDK which allows for the change of geometry effects. Merchant's original suggestion [8] was that it might be promising to explore the possibility of regarding the actual failure load λ_F as a function of some or all of these idealised load factors, ie,

$$\lambda = F\left(\lambda_C, \ \lambda_P, \ \lambda_y, \ \lambda_G,\right) \tag{14}$$

One particular concept that has been pursued, particularly by Murray in relation to triangulated frames [10,11], is to enquire whether there may not be some safe lower ratio of the failure load factor λ_F to the intersection load factor λ_G plotted in Fig 7. While λ_G is inevitably an upper bound on the value of λ_F , it can be argued that the essential components of behaviour are being allowed for - elastic response, plastic response and the effects of change of geometry. Murray fund that the ratio λ_F/λ_G lay between 0.77 and 0.98 for all the experimental results that he considered. However, for frame structures, Merchant suggested the empirical relationship

$$\frac{\lambda_F}{\lambda_P} + \frac{\lambda_F}{\lambda_c} = 1 \tag{15}$$

or, more revealingly, in the form

$$\lambda_F = \frac{\lambda_P}{1 + \lambda_P/\lambda_C} \tag{16}$$

This formula, which has become known as the Rankine-Merchant load, may be regarded as a generalization of Rankine's formula for the ultimate strength of struts. In that, for the great majority of frame structures in the practical range, the lowest elastic critical load factor λ_C is much greater than the rigid-plastic load factor λ_P equation (16) appears as a *modification* of the rigid-plastic load to allow for the effects of instability. For "stocky" frames, it gives $\lambda_F \approx \lambda_P$, thus reflecting the acknowledged usefulness of the simple rigid-plastic load in providing a reliable first approximation to the ultimate collapse loads of many practical frames.

3.4.2 Theoretical derivation of the Rankine-Merchant formula

Consider the four load-deflection relationships shown in Fig. 15. These are the linear elastic response Oaa' (ie. under *compensated* loading), the non-linear response Obb' (ie under *uncompensated* loading), the elastic-plastic response Oee' under *compensated* loading and the actual elastic-plastic response Ogg' (ie. under *uncompensated* loading). We postulate that, since the loading conditions for Obb' and Ogg' correspond respectively to those for Oaa' and Oee' (*uncompensated* in contrast to *compensated*), there is likely to be, at any particular deflection level Δ_1, a close equality between the respective load rations Δ_b/Δ_a to Δ_g/Δ_e. We will turn later to a discussion of the implications of making such an assumption.

We know that the two deflections Δ_1 and Δ_2 representing the two elastic responses at load level λ_b, are related to each other according to equation (9). If the loading produces a deflected state in the structure which is closely similar to that of the lowest elastic critical form (as is the case when a multi-storey frame is under sway loading), the first term in equation (9) dominates, whence

$$\Delta_1 = \frac{\Delta_2}{1 - \lambda_b/\lambda_c} \tag{17}$$

Now $\dfrac{\Delta_1}{\Delta_2} = \dfrac{\lambda_a}{\lambda_b}$ whence from (17)

$$\frac{\lambda_a}{\lambda_b} = \frac{\lambda_a + \lambda_c}{\lambda_c} = \frac{\Delta_1 + \Delta_c}{\Delta_c} \tag{18}$$

where Δ_c is the linear elastic deflection calculated at the lowest elastic critical load level λ_c as shown in Fig. 15.

The above result establishes the graphical construction shown in Fig 15 for determining the *uncompensated* load levels λ_b and λ_g at deflection level Δ_1 from the *compensated* load levels λ_a and λ_e respectively. In particular, by joining any point e at deflection Δ_1 on the *compensated* elastic-plastic response curve Oee' to the point h at $-\Delta_c$ on the deflection axis, the corresponding load level at deflection Δ_1 on the *uncompensated* elastic-plastic response curve is given by the intersection of this line he with the load level axis. In this way the complete *uncompensated* load-deflection curve may be derived from the *compensated* elastic-plastic curve.

Using the above result, a method of determining the elastic-plastic failure load from the *compensated* elastic-plastic curve immediately follows, as shown in Fig 16. The failure load factor λ_F is the intersection of the load factor axis of the tangent HDJ drawn to this curve from the deflection point $-\Delta_c$ on the deflection axis.

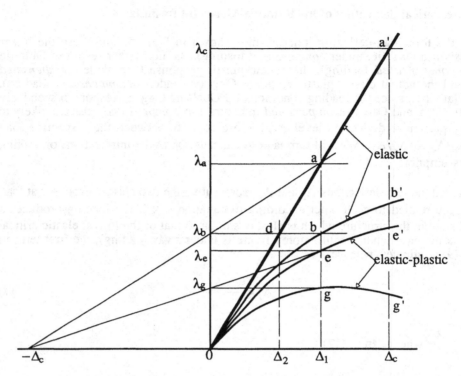

Figure 15   **Relationships between deflections caused by *compensated* (Oa', Oe')
and *uncompensated* (Ob', Og') loading.**

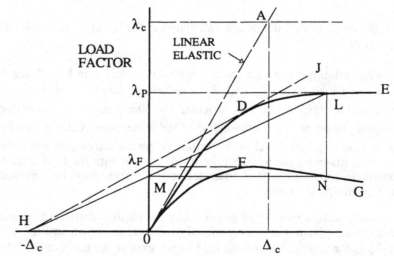

Figure 16   **Graphical construction for elastic-plastic failure load.**

An upper bound to the *compensated* elastic-plastic response is given by the relationship ODE in Fig 17, which follows the linear elastic response up to the rigid-plastic failure load factor λ_P, thereafter becoming horizontal. The application of the graphical procedure then gives the *uncompensated* response OFG, establishing a value for the failure load factor λ_F as shown.. (It may be noted that the idealized *compensated* load-deflection relation ODE is that which would be obtained if all the members had 'unit form-factors', that is full plastic moments of resistance equal to first yield moments, and if all the plastic hinges required to form a mechanism occurred simultaneously with the attainment of the rigid-plastic collapse load.)

It follows from similar triangles in Fig 17 that

$$\frac{\lambda_F}{\lambda_P} = \frac{\Delta_c}{\Delta_F + \Delta_c} \tag{19}$$

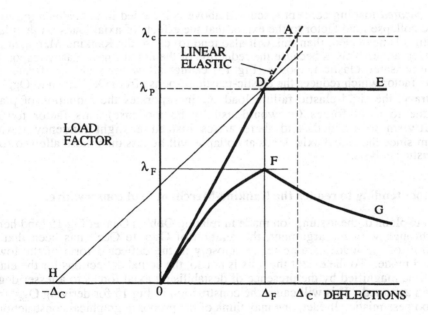

Figure 17 Graphical construction showing the significance of the Rankine-Merchant load.

By proportion, $\dfrac{\Delta_F}{\Delta_c} = \dfrac{\lambda_P}{\lambda_c}$, whence the Rankine-Merchant relation given by equation (16) follows immediately. This demonstration of the significance of the Rankine-Merchant load is due to Horne [13].

3.4.3   Limitations of the Rankine-Merchant load

By considering the various stages which have led to the above 'derivation' of the Rankine-Merchant load, it is possible to recognise more clearly the conditions under which it may be expected to provide a reasonably close estimate of the real failure load, and conversely which conditions tend to make it either a conservative or an unconservative estimate.

We consider in turn how various factors influence such tendencies.

3.4.4   Factors tending to render the Rankine-Merchant load unconservative

The main factor tending towards an optimistic prediction of the failure load is when there is a gradual yielding of the material over extensive lengths of the members so that wide-spread plastic zones occur well before the formation of complete plastic hinges, or when, because of high local bending moments, the first plastic hinges form at comparatively low load levels. These factors cause the elastic-plastic compensated response ODE in Fig 16 to fall well below the doubly linear relation ODE in Fig 17.

If the *compensated* loading concept discussed above is included in the calculation of the rigid-plastic collapse load factor to the extent that the effects of axial loads on the plastic moment values is neglected, then this will also tend to cause the Rankine-Merchant load to be unconservative. This is because the reduction of the upper linear elastic response to the non-compensated elastic response (Fig 15) cannot allow for yielding effects, thus ignoring one factor which reduces the elastic-plastic response from Oee' down to Ogg'. If, to the contrary, the rigid-plastic failure load λ_p incorporates the reduction of plastic moments due to axial forces (as would usually be the case), this factor towards unconservativism is avoided, and there arises instead a slight tendency towards conservatism since the actual axial loads at collapse will be less than those allowed for in the rigid-plastic analysis.

3.4.5   Factors tending to render the Rankine-Merchant load conservative.

As has been explained, the assumption made in relating Obb' to Oaa' in Fig 15 (and hence, following through with the argument, the relation of Ogg' to Oee') has been that the deflected form of the loaded frame relates closely to the deflected form of the lowest elastic critical mode. To the extent that this is not so, the actual deflections for the elastic frame would be magnified by the presence of destabilising axial forces to a lesser degree than has been assumed. This will cause the construction in Fig 15 for deriving Ogg' from Oee' to be too pessimistic. In fact, one may think of the proposed graphical construction as having to be modified by moving point H in Fig 16 further in the negative direction corresponding to the higher values of the linear elastic deflection Δ_c corresponding to the more relevant higher critical load levels. This effect is particularly obvious for the case of a multi-storey frame subjected to vertical loading only, since the rigid-plastic failure mode would be that of local beam failure, whereas the lowest elastic critical mode would be that of sway deformation of the frame as a whole.

The other main factors tending to render the Rankine-Merchant load conservative are related to the assumptions made about material behaviour. It has been assumed that indefinitely large strains can take place when the yield stress is reached, as represented by the idealised stress-strain relationship depicted in Fig 1(b). In accepting the concept of plastic hinge behaviour, it is assumed that infinitely large strains can take place without any rise of stress above the yield value. In practice, mild steel shows a rising stress-strain relationship (referred to as strain-hardening) after a finite strain (Fig 1 (a)), so that the moments at plastic hinge positions rise above the simply calculated full plastic values. Even in the absence of a rising stress-strain characteristic after yield, three-dimensional effects would restrict the capacity for high strains to occur at a theoretical plastic hinge position, causing the moments of resistance there to rise and make the rigid-plastic collapse load conservative.

To some extent these effects leading to conservative and unconservative estimates of the true failure load tend to compensate. We turn next to consider to what extent both theoretical and experimental evidence supports these conclusions.

3.5   EXAMPLES OF THE APPLICATION OF THE RANKINE-MERCHANT LOAD

3.5.1   Theoretical investigations

In his original paper, Merchant *et al* [8] reported on comparisons between the Rankine-Merchant loads of a large number of single bay, one- and two-storey frames and the actual theoretical failure loads of these same frames according to full elastic-plastic analyses (assuming unit form factor, ie. full plastic moment equal to first yield moment) carried out by Salem [12]. Salem's results are shown in Fig 18, where each dot represents a particular frame particularly loaded (see insets), the ratio of failure load to plastic collapse load being plotted vertically and the ratio of failure load to the least elastic critical load plotted horizontally. The straight line AB represents the Rankine-Merchant load calculated in accordance with equation (15). While there is (as would be expected) considerable scattering of results, there is certainly some evidence of a tendency for many results to crowd around in the vicinity of the Rankine-Merchant line. In view of the interplay and intricacy of many factors affecting the detailed behaviour of the frames, Merchant felt justified in claiming some significance for the value of his empirical approach. While some results lie below AB, the maximum shortfall is only of the order of 4%, thus helping to confirm the validity of the theoretical justification of the Rankine-Merchant load given in section 3.4.2.

The application of the Rankine-Merchant formula to the collapse loads of single members provides another opportunity for testing its validity where the material properties are still assumed to be purely elastic-plastic as in Fig 1b. When dealing analytically with the theoretical failure loads of eccentrically loaded struts, one is not dealing with a situation where a reasonably close analysis can be performed on the basis of a 'unit form-factor' treatment, with plasticity assumed to be confined to plastic hinge positions. Instead, one finds that plasticity spreads in zones not embracing the depth of the member, but comprising the stiffness to such a degree that very much reduced 'deteriorated critical loads' can cause failure long before any cross-section becomes fully plastic. Under these circumstances, one would expect the failure load to fall significantly below the Rankine-Merchant load estimate and this indeed proves to be the case.

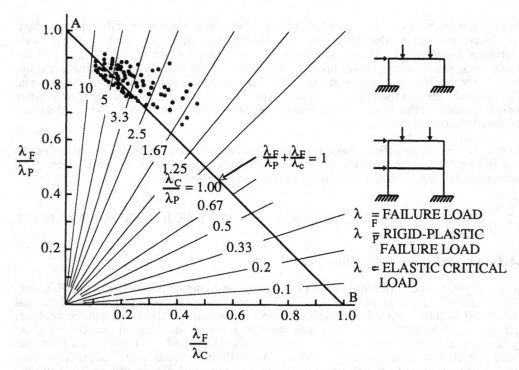

Figure 18   Theoretical failure load for one and two storey frames.

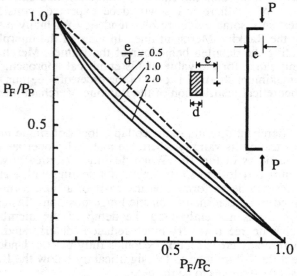

Figure 19   Comparison of theoretical collapse loads of eccentrically loaded struts with corresponding Rankine-Merchant Loads.

Fig 19 shows results obtained by Horne [13] for eccentrically loaded struts of rectangular cross-section, the curves corresponding to various loading eccentricities. They fall appreciably below the Rankine-Merchant load, shown dotted. The full range of member slenderness is covered in this theoretical treatment, low ratios of P_P/P_C, to the left of the diagram, corresponding to stocky struts, high ratios to slender struts.

3.5.2 Experimental investigations

Solely theoretical investigations of the validity for practical purposes of the Rankine-Merchant load suffer from the disadvantage that it is not easy to allow adequately for some of the factors mentioned in 3.4.5 as tending to make it conservative. Low [14] carried out tests to failure on over thirty model 3, 5 and 7 storey, single-bay frames, all constructed out of mild steel bars of square cross-section. His results are plotted in Fig 20. The ratio of experimental failure load to theoretical plastic collapse load is plotted vertically and of experimental failure load to lowest elastic critical load horizontally. In all cases, beams were equally loaded each by two equal loads at quarter span distances from the ends. Except for thirteen of the frames, equal side loads were also applied at each beam level. The results for those frames where no side loads were applied are distinguished by being encircled.

It will be seen that none of the experimental results lie below the Rankine-Merchant line AB. Moreover, in confirmation of the conclusion stated in section 3.4.5, the experimental failure loads for frames subjected solely to vertical loading lie well above AB.

3.5.3 Modification of the Rankine-Merchant load to allow for strain-hardening

Because of strain-hardening, which tends to raise plastic hinge values above those calculated from the yield stress, most practical frames of only a few storeys in height attain, at collapse, a load at least equal to the theoretical rigid-plastic collapse load. This is because the ratio λ_C/λ_P, the elastic critical load factor divided by the rigid-plastic collapse factor, indicated by the radial lines in Fig 20, is commonly greater than 10, so that the residual stiffness due to strain-hardening more than compensates for the effects of change of geometry. In addition to the effect of strain-hardening, it is found that even a small amount of stiffness from cladding is sufficient to compensate for the deterioration of stability due to plasticity. To allow in design for the minimum beneficial effects to be expected from strain-hardening and cladding, Wood [15] suggested a modification of the Rankine-Merchant relationship from AB in Fig 20 to ACB, ie,

$$\lambda_R^* = \frac{\lambda_P}{0.9 + \lambda_P/\lambda_C} \quad \text{when} \quad \frac{\lambda_C}{\lambda_P} < 10$$

$$\lambda_R^* = \lambda_P \quad \text{when} \quad \frac{\lambda_C}{\lambda_P} \geq 10 \tag{20}$$

Wood also proposed that, in order to avoid any difficulties that might arise due to instability effects in particularly slender frames, the above treatment should not be used when $\lambda_C/\lambda_P < 4$. His full proposal for application in practical design is therefore represented by ACD in Fig 20. He recommended that, when $\lambda_C/\lambda_P < 4$, a full analysis allowing fully for the effects of plasticity and instability should be carried out.

3.5.4   Approximate determination of elastic critical loads

Finally, in relation to the practical use in design of the Rankine-Merchant formula, including Wood's form, mention may be made of a convenient means of estimating the elastic critical loads of multi-storey rigid frames without having to use the facility for calculating critical loads available only in the more advanced computer packages. Since the influence of the critical load level in the practical range is merely to modify the simple plastic collapse load, equation (16), rather than being the *main* influence in determining the collapse load, an approximate estimate is certainly sufficient, say to within about 10%. The following procedure may be used to estimate the lowest critical load, Horne [16].

The *elastic critical load factor* to be calculated is that relevant to the system of loads, at unit load factor, for which an estimate of the *collapse load factor* is required. A linear elastic analysis is performed for the frame, with a *horizontal* load applied at each storey equal to the *vertical* load applied (at unit load factor) at that storey level. Then, if ϕ_i is the maximum sway index (lateral storey deflection divided by storey height) anywhere in the frame, the lowest elastic critical load factor is given by $0.9/\phi_i$.

An investigation [16] of a number of frames showed that this estimate always lay within less than 10% of the accurately calculated value.

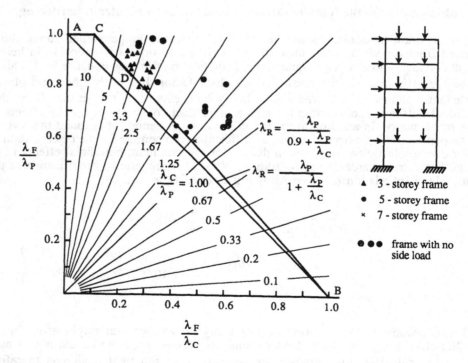

Figure 20 Experimental failure loads of model 3, 5 and 7 storey frames and modification of Rankine-Merchant load to allow for strain-hardening.

REFERENCES

1   Horne M R, "Fundamental propositions in the plastic theory of structures", J Inst Civ Engrs, vol 34 1949, p174

2   Greenberg H J and Prager W, "On limit design of beams and frames", Trans Amer Soc Civ Engrs, vol 117 1952, p447

3   Baker J F, Horne M R and Heyman J, "The Steel Skeleton vol 2", CUP 1956

4   Neal B G, "The plastic Methods of Structural Analysis", Chapman and Hall 1956

5   Hodge P G, Plastic Analysis of Structures, McGraw-Hill 1959

6   Horne M R and Morris J, Plastic Design of Low-rise Frames, Granada 1981

7   Horne M R, Plastic Theory of Structures, Permagon 1979

8   Merchant W, Rashid C A, Bolton A and Salem A, "The behaviour of unclad frames", Proc 50th Anniv Conf, Inst Struct Engrs, 1958

9   Wood R H, "The stability of tall buildings", Proc Inst Civ Engrs, vol 11 1978, p69

10 Murray N W, "The determination of the collapse loads of rigidly jointed frameworks with members in which axial forces are large", Proc Inst Civ Engrs Part III, vol 5 1956, p213

11 Murray N W, "Further tests on braced frameworks", Proc Inst Civ Engrs, vol 10 1958, p503

12 Salem A, Structural Frameworks, PhD Thesis, Univ Manchester 1958

13 Horne M R, "Elastic-plastic failure loads of plane frames", Proc Roy Soc A, vol 274 1963, p343

14 Low M W, "Some model tests on multi-storey rigid steel frames", Proc Inst Civ Engrs, vol 13 1959, p287

15 Wood R H, "Effective lengths of columns in multi-storey buildings", Structural Engineer, vol 52 1974, pp235, 295 & 341

16 Horne M R, "An approximate method for calculating the elastic critical loads of multi-storey plane frames", Structural Engineer, vol 53 1975, p242

THE SOUTHWELL AND THE DUNKERLEY THEOREMS

T. Tarnai
Technical University of Budapest, Budapest, Hungary

ABSTRACT

Summation formulae are used in the theory of elastic stability so that approximate estimates of the critical load factors of a complex problem are obtained by combining the load factors of subproblems in different ways. If the critical load factors are directly added, then the formula is called a *Southwell* type formula. If the reciprocals of the critical load factors are added, then the formula is called a *Dunkerley* type formula. The practical advantage of the summation formulae is that, for the subproblems, there are usually solutions available, or it is easy to determine them while, for the original problem the solution would be difficult to obtain. In this paper we will present the main theorems and formulae (Southwell theorem, Dunkerley theorem, Föppl-Papkovich theorem, Kollár conjecture, Melan theorem), and show, where possible, the conditions under which the results are on the safe side. The proofs are given in mathematical way, based on the relationships for eigenvalues of linear operators in Hilbert space. The theoretical results are illustrated by examples.

1. ON THE SUMMATION THEOREMS

First consider examples of the Southwell and Dunkerley formulae.

1.1 *Example*. Consider the torsional buckling of a thin-walled bar of height H with the lower end built in and the upper end free, subjected to a concentrated vertical force P at the free end. The equilibrium of the deformed bar can be described by the differential equation

$$EI_\omega \varphi'''' + \left(Pi_p^2 - GI_t \right) \varphi'' = 0 \tag{1.1}$$

with the boundary conditions

$$\varphi(H) = 0, \quad \varphi'(H) = 0, \quad \varphi''(0) = 0, \quad GI_t \varphi'(0) - EI_\omega \varphi'''(0) = Pi_p^2 \varphi'(0). \tag{1.2}$$

Here φ is the rotation of the cross-section, i_p is the radius of gyration. Prime denotes differentiation along the axis of the bar. The torsional stiffness of the cross-section is composed from two parts: the warping stiffness EI_ω and the *Saint-Venant* torsional stiffness GI_t. Let us consider first the case where the bar has warping stiffness only ($GI_t = 0$). In this case the critical load is [1]:

$$P_{cr,1} = \frac{1}{i_p^2} \frac{\pi^2 EI_\omega}{4H^2}.$$

Let us consider now the case where the bar has *Saint-Venant* torsional stiffness only ($EI_\omega = 0$). In this case the critical load is

$$P_{cr,2} = \frac{GI_t}{i_p^2}.$$

The *Southwell* summation approximates the critical load of the original bar by the expression

$$P_S = P_{cr,1} + P_{cr,2}. \tag{1.3}$$

The exact value of P_{cr} is given by [1]:

$$P_{cr} = \frac{1}{i_p^2} \left(GI_t + \frac{\pi^2 EI_\omega}{4H^2} \right).$$

It can be seen that in this example $P_S = P_{cr}$, that is, the *Southwell* formula (1.3) provides the exact value of the critical load.

1.2 *Example*. Fig. 1.1 shows a simply supported thin-walled bar of doubly symmetric ($I_x > I_y$) open cross-section, subjected to an axial force N and a couple M whose plane is perpendicular to the x axis. We want to determine the values of the N, M pairs by the *Dunkerley* formula under which the bar comes into the critical state. Consider first the case where the bar is subject to the force N only ($M=0$). Suppose that the critical load for torsional buckling is greater than the critical load for flexural buckling about the y axis. In this case the bar buckles in the weak direction, that is, about the y axis, and the *Euler* critical load is

$$N_{y,cr} = \frac{\pi^2 EI_y}{l^2}.$$

Let us consider now the case where the bar is subjected to the couple M only ($N=0$). The critical value of the couple can be obtained from [1]:

$$M_{cr} = \pm \sqrt{\frac{\pi^2 EI_y}{l^2}} \sqrt{\left(GI_t + \frac{\pi^2 EI_\omega}{l^2} \right)}.$$

Since N and M are of different dimension , the *Dunkerley* formula should be written in dimensionless form as follows:

$$\frac{N}{N_{y,cr}} + \frac{M}{M_{cr}} = 1 \qquad (1.4)$$

where M and M_{cr} should be considered with the same sign. To assess the accuracy of the *Dunkerley* formula (1.4), we have to determine the exact relationship between the critical pair N, M by solving the set of differential equations

$$EI_y u'''' + M\varphi'' + Nu'' = 0,$$

$$EI_\omega \varphi'''' - GI_t \varphi'' + Mu'' + Ni_p^2 \varphi'' = 0$$

with the boundary conditions

$$u(0) = u(l) = u''(0) = u''(l) = \varphi(0) = \varphi(l) = \varphi''(0) = \varphi''(l) = 0$$

where u is the displacement of the centroid of the cross-section in the y direction. If $M_{\varphi,cr}$ denotes the critical load of the bar for pure torsional buckling then the desired expression has the form

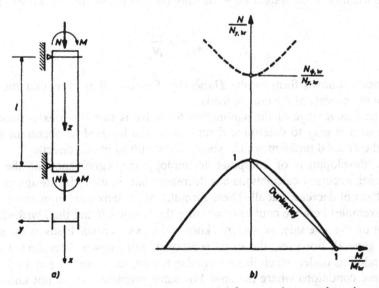

Fig.1.1 Lateral buckling of a bar subjected to an axial force and a couple at its ends. (a) The layout of the loads, (b) diagram of the loads and its approximation by the *Dunkerley* line

$$\left(1-\frac{N}{N_{y,cr}}\right)\left(1-\frac{N}{N_{\varphi,cr}}\right)-\frac{M^2}{M_{cr}^2}=0 \tag{1.5}$$

where [1]

$$N_{\varphi,cr}=\frac{1}{i_p^2}\left(GI_t+\frac{\pi^2 EI_\omega}{l^2}\right).$$

Let $N \le N_{y,cr} < N_{\varphi,cr} < \infty$, then in the co-ordinate system M/M_{cr}, $N/N_{y,cr}$, equation (1.4) represents a straight line and equation (1.5) represents a hyperbola, only one of whose branches should be considered (Fig. 1.1b). In the interval $0 < M/M_{cr} < 1$, points of the *Dunkerley* line (1.4) lie below the curve (1.5) inside the stability domain. Therefore the *Dunkerley* formula is on the safe side for any value of M/N but subject to the constraint $N > 0$. If N and M denote the exact values of the normal force and bending moment in a critical state, then instead of (1.4) the inequality below will be valid

$$\frac{N}{N_{y,cr}}+\frac{M}{M_{cr}}\ge 1. \tag{1.6}$$

If $N_{\varphi,cr} = \infty$, then the hyperbola is transformed into a parabola and the *Dunkerley* formula conservatively approximates with a very large error. If $N_{\varphi,cr} = N_{y,cr}$, then the hyperbola degenerates into two crossing straight lines and the *Dunkerley* formula is exact. [The case $N_{\varphi,cr} < N_{y,cr}$ was excluded.]

When the ratio of M and N is fixed, for instance, force N acts with an excentricity e ($M=eN$), the loading force system depends only on one parameter: N. From (1.6) for N_{cr} we obtain

$$\frac{1}{N_{y,cr}}+\frac{1}{\dfrac{M_{cr}}{e}}\ge\frac{1}{N_{cr}} \tag{1.7}$$

which is a better known form of the *Dunkerley* formula. It is seen that the summation applies to the reciprocals of the critical loads.

The practical advantage of the summation formulae is that there exist solutions to the subproblems or it is easy to determine them, which can be used to obtain an approximate solution for the original problem which would be difficult to obtain directly.

With the development of computer technology the significance of the summation formulae in high accuracy calculations has decreased but, in everyday design practice, their significance has not decreased at all. These formulae are widely used in practice.

In both examples here we could say whether the *Southwell* and the *Dunkerley* formulae are or are not on the safe side, as we have known the exact critical loads of the structures in question. In general, however, the exact solution is not known. Therefore it is useful to know the conditions under which these formulae are conservative. The aim of this paper is to present these conditions where possible. For some problems we do not know them; and for some types of problems *no such conditions exist*.

The algebraic forms of the summation formulae are very simple, although they are often difficult to derive. The application of functional analysis (theory of Hilbert spaces and linear operators defined in them) can be effective in this respect. The mathematical base of these formulae can be found in [2] and [3]. Some results prepared for engineers are presented in [4].

Most of the mathematical terms were defined previously by Weinberger [3]. Here we introduce some additional ones.

Definition. The set X is dense in the set Y if any arbitrarily small neighbourhood of any element of Y contains elements of X. (The set of rational numbers is dense in the set of real numbers, for instance.)

Definition. Consider a linear operator S defined in a Hilbert space H. Let (u,v), u, $v \in H$ denote the scalar product in H, and D_S the domain of definition of S. We say that the linear operator S is

(i) symmetric, if $(Su, v) = (u, Sv)$ for all u, $v \in D_S$ and D_S is dense in H;

(ii) positive, if $(Su, u) > 0$ for all $u \neq 0$, $u \in D_S$;

(iii) positive definite, if there exists a number $\alpha > 0$ such that $(Su, u) \geq \alpha(u, u), u \in D_S$.

Definition. The spectrum of a linear operator A is the set of numbers λ such that the operator $A - \lambda I$ has no unique bounded inverse. (In a general case, the identity operator I is replaced by a linear operator B and we refer to the spectrum of an operator pair.) An operator (or operator pair) comprises a discrete spectrum if the spectrum consists only of eigenvalues.

Theorem. Let A and B be symmetric linear operators in H such that (Bu,u) is completely continuous with respect to (Au,u) (the definition of the term "completely continuous" is given by Weinberger [3]), then the operator pair $A - \lambda B$ is of discrete spectrum. (We do not prove this here.)

2. THE SOUTHWELL THEOREM

In the theory of structural stability the *Southwell* theorem is the following: If the stiffness of a structure is composed of parts, then the smallest critical load parameter of the structure is not less than the sum of the smallest critical load parameters corresponding to the partial stiffnesses. The theorem can be stated purely in mathematical terms as follows.

T h e o r e m 2.1 *Let A and B be symmetric linear operators in a Hilbert space H such that A is positive definite, B is positive and $D_A \subset D_B \subset H$. Suppose (Bu,u) is completely continuous with respect to (Au,u). Let $A = \sum_{i=1}^{n} A_i$ where the operators A_i $(i = 1, 2, ..., n)$ have the same properties as A. If λ_0 and λ_i denote the smallest eigenvalues of the eigenvalue problem*

$$(A - \lambda B)u = 0 \tag{2.1}$$

and

$$(A_i - \lambda B) = 0, \qquad i = 1, 2, ..., n, \tag{2.2}$$

respectively, then we have

$$\lambda_0 \geq \sum_{i=1}^{n} \lambda_i. \tag{2.3}$$

P r o o f . Because A_i is positive definite and B is positive the *Rayleigh* quotient is bounded from below and

$$\lambda_i = \inf_{u \in D_{Ai}} \frac{(A_i u, u)}{(Bu, u)}, \qquad i = 1, 2, ..., n. \tag{2.4}$$

Let u_0 denote the eigenelement corresponding to the eigenvalue λ_0. On the basis of (2.4) it follows that

$$\lambda_0 = \frac{(A u_0, u_0)}{(B u_0, u_0)} = \sum_{i=1}^{n} \frac{(A_i u_0, u_0)}{(B u_0, u_0)} \geq \sum_{i=1}^{n} \inf_{u \in D_{Ai}} \frac{(A_i u, u)}{(Bu, u)} = \sum_{i=1}^{n} \lambda_i.$$

That proves the theorem.

R e m a r k s . It is important in the theorem that B is positive. If B is not positive then the the eigenvalue problems may have also negative eigenvalues and so the statement in the theorem is not valid.

In the proof, in fact, complete continuity is not used, but the property that the functional $(A_i u, u) / (Bu, u)$ is bounded from below. The theorem is valid for a degenerate case where (Bu, u) is not completely continuous with respect to $(A_i u, u)$ but $(A_i u, u) / (Bu, u) > 0$, $u \in D_{Ai}$ and λ_i is defined as the infimum of the *Rayleigh* quotient $(A_i u, u) / (Bu, u)$. This occurs, for instance if, for some i the spectrum of (2.2) is only one point , but it is an eigenvalue with infinite multiplicity as occurs in Example 1.1 of Section 1.

2.1 *Example.* Consider a column of variable cross-section with the lower end built in and the upper end free. Let the column be subject to the action of a distributed axial load given by the equation $q = q_1 (l - z) / l$ (Fig. 2.1). Let the variation of the second moment of area of the cross-section be given by the equation

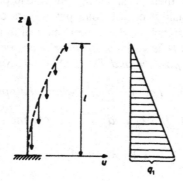

Fig. 2.1 Buckling of a cantiliver with variable cross-section

$$I = I_1 + I_2 \frac{l-z}{l} + I_3 \left(\frac{l-z}{l}\right)^2 + I_4 \left(\frac{l-z}{l}\right)^3.$$

The equilibrium of the bar can be described by the equation

$$\left(EIu''\right)'' + q_1 \left[\frac{(l-z)^2}{2l} u'\right]' = 0 \qquad (2.5)$$

with the boundary conditions

$$u(0) = u'(0) = u''(l) = u'''(l) = 0. \qquad (2.6)$$

Let us introduce the following notation:

$$A_1 u = \left(EI_1 u''\right)'', \qquad (2.7)$$

$$A_2 u = \left(EI_2 \frac{l-z}{l} u''\right)'', \qquad (2.8)$$

$$A_3 u = \left[EI_3 \left(\frac{l-z}{l}\right)^2 u''\right]'', \qquad (2.9)$$

$$A_4 u = \left[EI_4 \left(\frac{l-z}{l}\right)^3 u''\right]'', \qquad (2.10)$$

$$Au = \sum_{i=1}^{4} A_i u, \qquad (2.11)$$

$$Bu = -\left[\frac{(l-z)^2}{2l} u'\right]', \qquad (2.12)$$

$$\lambda = q_1. \qquad (2.13)$$

Differential expressions (2.7 to 12) with the boundary conditions (2.6) define differential operators satisfying the conditions of Theorem 2.1, and the eigenvalue problem (2.5), (2.6) may be written in the form $Au - \lambda Bu = 0$, and the eigenvalue problems $A_i u - \lambda Bu = 0$ ($i = 1,\dots, 4$) can be defined. If λ_0 denotes the least critical value of the load parameter of the original bar, and λ_i denotes the least critical load parameter of the bar with bending rigidity $EI_i \left[(l-z)/l\right]^{i-1}$, then from the *Southwell* Theorem 2.1 we have

$$\lambda_0 \ge \sum_{i=1}^{4} \lambda_i. \qquad (2.14)$$

The values λ_i ($i = 1,\dots,4$) can be determined by Tables 2-14 on page 131 of [1], and a lower bound of the desired value of the critical load parameter λ_0 obtained from (2.14).

As an aid to understanding it is useful to clarify the meaning of the conditions in Theorem 2.1 and show how they are satisfied.

We are looking for the least eigenvalue λ of equation (2.5) in the following class of functions u: u is four times continuously differentiable in the open interval $\{0,l\}$, three times continuously differentiable in the interval $\{0,l]$ open from left and closed from right, continuously differentiable in the closed interval $[0,l]$, and satisfies the boundary conditions (2.6).

Functions u are once continuously differentiable in $[0,l]$, so they are continuous and consequently bounded in $[0,l]$. From this it follows that they are square integrable in $[0,l]$. For an arbitrary pair u, v of such functions there exists the integral

$$(u,v) = \int_0^l uv\,dz$$

which has the properties of the scalar product. These functions are therefore elements of $L^2[0,l]$, that is, the space of square integrable functions in the interval $[0,l]$. The space $L^2[0,l]$ is a *Hilbert* space.

Consider the operator

$$A = \frac{d^2}{dz^2}\left(EI\frac{d^2}{dz^2}\right)$$

defined on the class of functions u. Because of the linear properties of differentiation, A is a linear operator. A is symmetric since for, arbitrary functions u, $v \in D_A$ with integration in parts we obtain

$$(Au,v) = \int_0^l (EIu'')'' v\,dz = \left[(EIu'')' v\right]_0^l - \left[EIu''v'\right]_0^l +$$

$$\left[u'EIv''\right]_0^l - \left[u(EIv'')'\right]_0^l + \int_0^l u(EIv'')''\,dz = (u,Av).$$

As a consequence of the boundary conditions (2.6) all the integrated parts are equal to zero. *In statics, symmetry of operator A means the equality of the internal "alien" works, that is, if there is a set of stress and a set of strain independent of each other, then their internal work is equal to the internal work of the set of strain and the set of stress corresponding to the given sets of stress and strain due to Hooke's law.*

D_A is dense in $L^2[0,l]$. It comes from the fact that any square integrable function in $[0,l]$ can be approximated almost everywhere in $[0,l]$ with arbitrary exactness by functions which are four times continuously differentiable in $\{0,l\}$, three times continuously differentiable in $\{0,l]$ and once continuously differentiable in $[0,l]$ (for instance, by polynomials).

It is also possible to show that the operator A is positive. Let $u \in D_A$, $u \neq 0$ be arbitrary then, integrating by parts, we obtain

$$(Au,u) = \int_0^l (EIu'')'' u\,dz = \left[(EIu'')' u\right]_0^l - \left[EIu''u'\right]_0^l + \int_0^l EIu''u''\,dz = \int_0^l EIu''^2\,dz > 0$$

because $EI > 0$ and the integrated parts are equal to zero in consequence of the boundary conditions (2.6). *In statics, positivity of operator A means the positivity of the strain energy.* (The scalar product is twice the strain energy.)

It is possible to show that A is positive definite but we omit this now.

Introduce the notation $g(z) = (l - z)^2 / (2l)$ by which operator B in (2.12) can be written as

$$B = -\frac{d}{dz}\left[g(z)\frac{d}{dz}\right].$$

This operator is defined on the class of functions twice continuously differentiable in the open interval $\{0,l\}$, once continuously differentiable in the interval $[0,l\}$ closed from left open from right, vanishing at point $z = 0$. As three times and four times differentiability is not required it follows that the domain of definition of B contains that of A, that is, $D_A \subset D_B$.

Operator B is linear because of the linear properties of differentiation. B is symmetric since for any $u, v \in D_B$ by integration by parts it is obtained:

$$(Bu, v) = -\int_0^l \left[g(z)u'\right]' v\,dz = -\left[g(z)u'v\right]_0^l + \left[ug(z)v'\right]_0^l - \int_0^l u\left[g(z)v'\right]'\,dz = (u, Bv).$$

The integrated parts are equal to zero because of the boundary conditions $u(0)=0$, $u'(0)=0$ in (2.6) and $g(l)=0$. *In statics, symmetry of operator B means the equality of the external "alien" works, that is, the validity of the Maxwell-Betti theorem.*

D_B is dense in $L^2[0,l]$. This can be shown similar to the case of operator A.

Operator B is positive because for any $u \in D_B$, $u \neq 0$ integration by parts yields:

$$(Bu, u) = -\int_0^l \left[g(z)u'\right]' u\,dz = -\left[g(z)u'u\right]_0^l + \int_0^l g(z)u'u'\,dz = \int_0^l g(z)u'^2\,dz > 0.$$

This holds as $g(z) \geq 0$, $z \in [0,l]$, and the integrated parts are equal to zero in consequence of the boundary conditions $u(0) = 0$, $u'(0) = 0$ in (2.6) and $g(l) = 0$. *In statics, positivity of operator B means the positivity of external work.* (The scalar product is twice the external work.)

To show that (Bu,u) is completely continuous with respect to (Au,u) is a larger task that we omit here. *In statics, complete continuity results in that the load has denumerable (infinitely many) critical values. These are well separated, and a definite buckling form belongs to each of them.*

3. DUNKERLEY TYPE THEOREMS AND FORMULAE

3.1 The Dunkerley theorem

In the theory of stability the classical *Dunkerley* theorem is the following: The reciprocal of the least critical load parameter of an elastic structure subjected to a complex load system is not greater than the sum of the reciprocals of the least critical load parameters of the same structure subjected to subsystems of the load. In this theorem mathematically it is supposed that all the operators are positive definite or positive, that is, where all the eigenvalues (critical load parameters) are positive. The theorem, however, can be generalized for cases where there are both positive and negative eigenvalues. Now we formulate the *Dunkerley* theorem mathematically for this generalised case.

Theorem 3.1 *Let A and B be symmetric linear operators in a Hilbert space H such that A is positive, and there exists a constant c such that*

$$|(Bu,u)| \leq c(Au,u), \quad u \in D_A \subset D_B \subset H. \tag{3.1}$$

Suppose that the quadratic functional (Bu,u) is completely continuous with respect to (Au,u). Let $B = \sum_{i=1}^{n} B_i$ *where the operators* B_i *(i = 1, 2,..., n) have the same properties as B, and let there exist constants* c_i *associated to* B_i *to satisfy (3.1). If* λ_0 *and* λ_i *denote the least positive eigenvalues of the eigenvalue problem*

$$(A - \lambda B)u = 0 \tag{3.2}$$

and

$$(A - \lambda B_i)u = 0, \quad i = 1, 2, ..., n, \tag{3.3}$$

respectively, then we have

$$\frac{1}{\lambda_0} \leq \sum_{i=1}^{n} \frac{1}{\lambda_i}. \tag{3.4}$$

Proof. Because of (3.1) the reciprocal of the *Rayleigh* quotient

$$(B_i u, u) / (Au, u) \quad i = 1, 2, ..., n$$

is bounded and

$$\frac{1}{\lambda_i} = \sup_{u \in D_A} \frac{(B_i u, u)}{(Au, u)}. \tag{3.5}$$

Let u_0 denote the eigenelement corresponding to eigenvalue λ_0. On the basis of (3.5) it follows that

$$\frac{1}{\lambda_0} = \frac{(Bu_0, u_0)}{(Au_0, u_0)} = \sum_{i=1}^{n} \frac{(B_i u_0, u_0)}{(Au_0, u_0)} \leq \sum_{i=1}^{n} \sup_{u \in D_A} \frac{(B_i u, u)}{(Au, u)} = \sum_{i=1}^{n} \frac{1}{\lambda_i}$$

that proves the theorem.

Remark. If for some i, $(A - \lambda B_i)u = 0$ has no positive eigenvalues then

$$\sup_{u \in D_A} \frac{(B_i u, u)}{(Au, u)} = 0,$$

and in (3.4), $1/\lambda_i$ should be replaced by 0.

In the following, some examples of the application of the *Dunkerley* theorem will be presented for lateral buckling of beam-columns. Equilibrium of thin-walled bars with symmetric open cross-section subjected to both axial and transverse forces in a buckled state can be described by the differential equations below [4] if the normal force N is constant:

$$EI_y u'''' + (M_x \varphi)'' + N(u'' + y_0 \varphi'') = 0, \tag{3.6}$$

$$EI_\omega \varphi'''' - GI_t \varphi'' + M_x u'' + tM_x'' \varphi - \beta_1(M_x \varphi')' + N(y_0 u'' + i_{pT}^2 \varphi'') = 0. \tag{3.7}$$

Here t is the distance to the point of application of the transverse load q from the shear centre, and β_1, y_0, i_{pT}^2 are cross-sectional constants. Suppose we have a one-parameter load system. Let λ denote the load parameter, M_{x1}, N_1, q_1 the reference values of the functions M_x, N, q. So,

$$M_x = \lambda M_{x1}, \quad N = \lambda N_1, \quad q = \lambda q_1.$$

Introduce the following notation

$$A_{11} u = EI_y u''''',$$ (3.8)

$$A_{22} \varphi = EI_\omega \varphi'''' - GI_t \varphi'',$$ (3.9)

$$B_{11} u = -N_1 u'',$$ (3.10)

$$B_{12}^{(1)} \varphi = -y_0 N_1 \varphi'',$$ (3.11)

$$B_{12}^{(2)} \varphi = -\left(M_{x1} \varphi\right)'',$$ (3.12)

$$B_{21}^{(1)} u = -y_0 N u'',$$ (3.13)

$$B_{21}^{(2)} u = -M_{x1} u'',$$ (3.14)

$$B_{22}^{(1)} \varphi = -i_{pT}^2 N_1 \varphi'',$$ (3.15)

$$B_{22}^{(2)} \varphi = \beta_1 \left(M_{x1} \varphi'\right)',$$ (3.16)

$$B_{22}^{(3)} \varphi = -t M_x'' \varphi,$$ (3.17)

$$B_{12} = B_{12}^{(1)} + B_{12}^{(2)},$$ (3.18)

$$B_{21} = B_{21}^{(1)} + B_{21}^{(2)},$$ (3.19)

$$B_{22} = B_{22}^{(1)} + B_{22}^{(2)} + B_{22}^{(3)}.$$ (3.20)

Differential expressions (3.8 to 17) with given homogeneous boundary conditions denote differential operators. In this way equations (3.6, 7) with given boundary conditions can be written in the form

$$\begin{bmatrix} A_{11} & 0 \\ 0 & A_{22} \end{bmatrix} \begin{bmatrix} u \\ \varphi \end{bmatrix} - \lambda \begin{bmatrix} B_{11} & B_{12} \\ B_{21} & B_{22} \end{bmatrix} \begin{bmatrix} u \\ \varphi \end{bmatrix} = \begin{bmatrix} 0 \\ 0 \end{bmatrix},$$ (3.21)

that has a shorter form $(A - \lambda B) w = 0$ by introducing notation

$$A = \begin{bmatrix} A_{11} & 0 \\ 0 & A_{22} \end{bmatrix}, \quad B = \begin{bmatrix} B_{11} & B_{12} \\ B_{21} & B_{22} \end{bmatrix}, \quad w = \begin{bmatrix} u \\ \varphi \end{bmatrix}.$$ (3.22)

In most practical cases the boundary conditions result in operators A and B satisfying the conditions of Theorem 3.1.

3.1.1 *Example.* Let operator B be of the form

$$B = B_1 + B_2$$ (3.23)

where

$$B_1 = \begin{bmatrix} B_{11} & B_{12}^{(1)} \\ B_{21}^{(1)} & B_{22}^{(1)} \end{bmatrix}, \quad B_2 = \begin{bmatrix} 0 & B_{12}^{(2)} \\ B_{21}^{(2)} & B_{22}^{(2)} + B_{22}^{(3)} \end{bmatrix}.$$ (3.24)

This means that the system of external forces is divided into two parts: (a) axial and (b) transverse loads (Fig. 3.1). Due to Theorem 3.1 a lower bound can be given to the least positive critical load parameter by the least critical load parameter for torsional buckling and the least positive critical load parameter for lateral buckling.. This is a traditional way of application of the *Dunkerley* theorem.

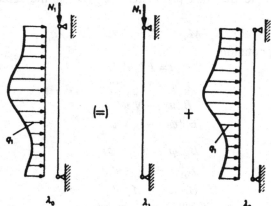

Fig. 3.1 Subdivision of the load into axial and transverse components.

3.1.2 *Example.* Consider the problem of lateral buckling of the bar without axial forces $(N_1 = 0)$:

$$\begin{bmatrix} A_{11} & 0 \\ 0 & A_{22} \end{bmatrix}\begin{bmatrix} u \\ \varphi \end{bmatrix} - \lambda \begin{bmatrix} 0 & B_{12}^{(2)} \\ B_{21}^{(2)} & B_{22}^{(2)} + B_{22}^{(3)} \end{bmatrix}\begin{bmatrix} u \\ \varphi \end{bmatrix} = \begin{bmatrix} 0 \\ 0 \end{bmatrix}. \tag{3.25}$$

Let us subdivide operator B_2 in (3.24) into two parts:

$$B_2 = B_2^{(1)} + B_2^{(2)} \tag{3.26}$$

such that

$$B_2^{(1)} = \begin{bmatrix} 0 & B_{12}^{(2)} \\ B_{21}^{(2)} & B_{22}^{(2)} \end{bmatrix}, \qquad B_2^{(2)} = \begin{bmatrix} 0 & 0 \\ 0 & B_{22}^{(3)} \end{bmatrix}. \tag{3.27}$$

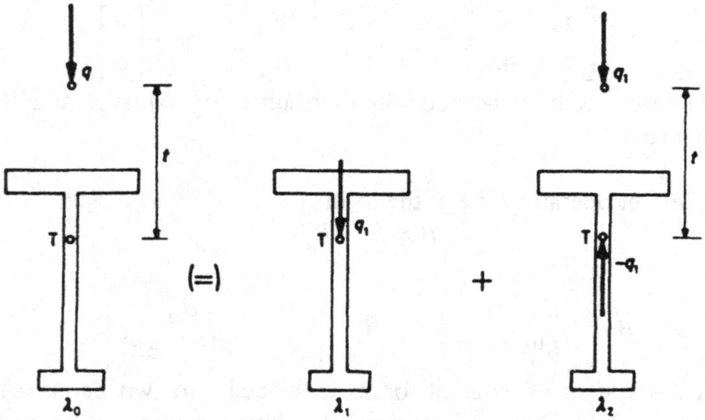

Fig. 3.2 Grouping of loads

Let operator $B_{22}^{(3)}$ in (3.17) be positive, that is, let the point of application of the load be above the shear centre. In this case the conditions of Theorem 3.1 are satisfied by the operators A, B_2, $B_2^{(1)}$, $B_2^{(2)}$ defined by (3.22), (3.26), (3.27), and to the least positive critical load parameter we can give an approximate value in the following way (Fig. 3.2): First we determine the least positive critical load parameter for the case where the point of application of the load is the shear centre $\left(B_2^{(2)} = 0\right)$. This is λ_1. Then we determine the critical load of the structure for the case where additionally to the original load with original point of application, another load is applied whose intensity is the same as that of the original, but its direction is the opposite, and its point of application is the shear centre $\left(B_2^{(1)} = 0\right)$. This is equivalent to the case where the bar is simply supported along an axis passing through the shear centre. This support allows the cross-sections to rotate. The critical load parameter in this case is λ_2. The theorem yields the inequality

$$\frac{1}{\lambda_0} \leq \frac{1}{\lambda_1} + \frac{1}{\lambda_2}.$$

3.1.3 *Example.* Another way of solving the problem (3.25) can be if operator B_2 is divided into two parts as

$$B_2 = \tilde{B}_2^{(1)} + \tilde{B}_2^{(2)} \tag{3.28}$$

such that

$$\tilde{B}_2^{(1)} = \begin{bmatrix} 0 & B_{12}^{(2)} \\ B_{21}^{(2)} & B_{22}^{(3)} \end{bmatrix}, \quad \tilde{B}_2^{(2)} = \begin{bmatrix} 0 & 0 \\ 0 & B_{22}^{(2)} \end{bmatrix}. \tag{3.29}$$

Operators A, B_2, $\tilde{B}_2^{(1)}$, $\tilde{B}_2^{(2)}$ defined by (3.22), (3.28), (3.29) satisfy the conditions of Theorem 3.1, therefore

$$\frac{1}{\lambda_0} \leq \frac{1}{\lambda_1} + \frac{1}{\lambda_2},$$

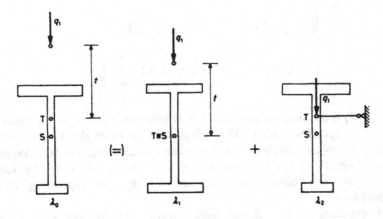

Fig. 3.3 Imaginary transformation of the structure for the *Dunkerley* theorem

where λ_1 is the least positive critical load parameter of such a bar whose cross-section is doubly symmetric and whose stiffness properties are the same as those of the original bar, and where the distance between the point of application of the transverse load and the shear centre is the same as that of the original bar; λ_2 is the least positive critical load parameter of the original bar subjected to the load at the shear centre ($t = 0$) when the lateral displacement of the shear centre is prevented (Fig. 3.3).

3.1.4 *The Strigl formula.* There are known investigations to sharpen the *Dunkerley* inequality (3.4) in the Theorem 3.1, that is, to bring the inequality closer to equality (Fig. 3.4). These investigations led to the non-linear versions of the *Dunkerley* formula.

Let A and B symmetric linear operators in a *Hilbert* space H such that A is positive definite and B is positive. In order to have a discrete spectrum, let us suppose that the quadratic functional (Bu,u) is completely continuous with respect to (Au,u). Let $B = B_1 + B_2$ where the operators B_1 and B_2 have the same properties as B. If λ_0 denotes the least eigenvalue of the eigenvalue problem

$$(A - \lambda B)u = 0, \tag{3.30}$$

and λ_1, u_1 and λ_2, u_2 denote the least eigenvalues and the corresponding eigenelements of the eigenvalue problems

$$(A - \lambda B_1)u = 0 \tag{3.31}$$

and

$$(A - \lambda B_2)u = 0, \tag{3.32}$$

respectively, then according to *Strigl* [5], a very good approximation is provided by the quadratic *Dunkerley* formula

$$\frac{\lambda_0}{\lambda_1} + \frac{\lambda_0}{\lambda_2} - \frac{\lambda_0^2}{\lambda_1 \lambda_2}\left[1 - \frac{(Au_1, u_2)^2}{(Au_1, u_1)(Au_2, u_2)}\right] \approx 1 \tag{3.33}$$

which can be written in the form

$$\begin{vmatrix} (Au_1, u_1)\left(1 - \dfrac{\lambda_1}{\lambda_0}\right) & (Au_1, u_2) \\[2ex] (Au_1, u_2) & (Au_2, u_2)\left(1 - \dfrac{\lambda_2}{\lambda_0}\right) \end{vmatrix} \approx 0. \tag{3.34}$$

Relationship (3.34) is the simplest case of a highly non-linear interaction formula derived by *Strigl* [5]. He has applied formula (3.33) with success for cases also where operator B was not positive but only symmetric, and so the problem had also negative eigenvalues. Practical application of formula (3.33) is somewhat difficult since not only the least positive eigenvalues of the subproblems are required but the corresponding eigenelements (eigenfunctions) also.

Numerical experiments of *Strigl* have shown that formula (3.33) can lead to approximations both on the safe side and the unsafe side (Fig. 3.4*b*).

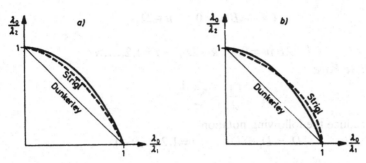

Fig. 3.4 Approximation more exact than the *Dunkerley* straight line: (a) on the safe side,
(b) partly on the safe side, partly on the unsafe side.

3.2 The Föppl-Papkovich theorem

The algebraic form of the *Föppl-Papkovich* formula is the same as that of the *Dunkerley* formula, but its physical background is different. Its essence is the following. Let a structure be characterised by n stiffness parameters. Consider an imaginary structure obtained from the original, all of whose stiffness parameters are increased up to infinity except the ith one which is kept unchanged. Let the least critical load parameter of the structure obtained in this way be denoted by λ_i. Let this procedure be done for $i = 1, 2, \dots, n$. If λ_0 denotes the least critical load parameter of the original structure, then for its reciprocal a good approximation is given by the reciprocals of the least critical load parameters of the imaginary rigidized structures:

$$\frac{1}{\lambda_0} \approx \sum_{i=1}^{n} \frac{1}{\lambda_i}. \tag{3.35}$$

However, it is not clear whether or not the approximation is conservative. In the following we will give conditions under which formula (3.35) results in approximation on the safe side.

Definition. Let A be a symmetric linear operator in *Hilbert* space H. We say that elements u and v $(u, v \in D_A)$ are A-orthogonal, if $(A u, v) = 0$. Let X and Y be subspaces in D_A. We say that X and Y are A-orthogonal, if for every $u \in X$ and $v \in Y$ we have $(Au, v) = 0$. Let Z be a subspace in D_A. We say that Z is A-orthogonal direct sum of X and Y, denoted by $Z = X \oplus Y$, if every $w \in Z$ can be written in the form $w = u + v$ such that $u \in X$, $v \in Y$ and $(Au, v) = 0$.

Theorem 3.2 *Let A and B be symmetric linear operators in Hilbert space H such that A is positive definite, B is positive and $D_A \subset D_B \subset H$. Let the quadratic functional (Bu, u) be completely continuous with respect to (Au, u). Let $D_1, D_2, \dots, D_n$ be pairwise A-orthogonal subspaces in D_A such that $D_A = D_1 \oplus D_2 \oplus \dots \oplus D_n$. If λ_0 and λ_i denotes the least eigenvalue of the eigenvalue problem*

$$(A - \lambda B)u = 0, \quad u \in D_A \tag{3.36}$$

and

$$(A - \lambda B)u = 0, \quad u \in D_i, \quad i = 1, 2, ..., n, \tag{3.37}$$

respectively, then we have

$$\frac{1}{\lambda_0} \leq \sum_{i=1}^{n} \frac{1}{\lambda_i}. \tag{3.38}$$

P r o o f . Introduce the following notation

$$\tilde{D}_i = D_i \oplus \tilde{D}_{i+1}, \quad i = 1, 2, ..., n-1 \tag{3.39}$$

where

$$\tilde{D}_1 = D_A, \quad \tilde{D}_n = D_n. \tag{3.40}$$

As D_i is a subspace, it follows that $\tilde{D}_i$ $(i = 1, 2, ..., n)$ is also a subspace in D_A. Let $\tilde{\lambda}_i$ denote the least eigenvalue of the eigenvalue problem

$$(A - \lambda B)u = 0, \quad u \in \tilde{D}_i, \quad i = 1, 2, ..., n \tag{3.41}$$

where

$$\tilde{\lambda}_1 = \lambda_0 \quad \text{and} \quad \tilde{\lambda}_n = \lambda_n. \tag{3.42}$$

Let us fix the value of i. Due to (3.39), $\tilde{D}_i$ is composed as a direct sum of two A-orthogonal subspaces: $\tilde{D}_i = D_i \oplus \tilde{D}_{i+1}$, that is , every $u \in \tilde{D}_i$ has a decomposition of the form

$$u = v + w$$

such that $v \in D_i$, $w \in \tilde{D}_{i+1}$ and $(Av, w) = 0$. Because A is positive definite and B is positive, the eigenvalues $\tilde{\lambda}_i, \lambda_i, \tilde{\lambda}_{i+1}$ can be given by the *Rayleigh* quotient:

$$\frac{1}{\tilde{\lambda}_i} = \sup_{u \in \tilde{D}_i} \frac{(Bu, u)}{(Au, u)}, \tag{3.43}$$

$$\frac{1}{\lambda_i} = \sup_{v \in D_i} \frac{(Bv, v)}{(Av, v)}, \tag{3.44}$$

$$\frac{1}{\tilde{\lambda}_{i+1}} = \sup_{w \in \tilde{D}_{i+1}} \frac{(Bw, w)}{(Aw, w)}. \tag{3.45}$$

From (3.44) and (3.45) it follows that

$$(Bv, v) \leq \frac{1}{\lambda_i}(Av, v), \quad v \in D_i, \tag{3.46}$$

$$(Bw, w) \leq \frac{1}{\tilde{\lambda}_{i+1}}(Aw, w), \quad w \in \tilde{D}_{i+1}. \tag{3.47}$$

B is a positive operator, therefore the scalar product (Bu, u) defines a norm on D_B:

$$\|u\|_B = (Bu, u)^{1/2}, \tag{3.48}$$

for which the triangle inequality holds, that is, for $u = v + w$

$$\|u\|_B = \|v + w\|_B \leq \|v\|_B + \|w\|_B. \tag{3.49}$$

From (3.48), (3.49) and (3.46), (3.47) it follows that if $u = v + w$, $v \in D_i$, $w \in \tilde{D}_{i+1}$ then

$$(Bu,u) = (B(v+w), v+w) = \|v+w\|_B^2 \le \left[\|v\|_B + \|w\|_B\right]^2 = \left[(Bv,v)^{1/2} + (Bw,w)^{1/2}\right]^2 \le$$

$$\left\{\left[\frac{1}{\lambda_i}(Av,v)\right]^{1/2} + \left[\frac{1}{\tilde{\lambda}_{i+1}}(Aw,w)\right]^{1/2}\right\}^2 =$$

$$\frac{1}{\lambda_i}(Av,v) + \frac{1}{\tilde{\lambda}_{i+1}}(Aw,w) + 2\left[\frac{1}{\lambda_i}(Av,v)\frac{1}{\tilde{\lambda}_{i+1}}(Aw,w)\right]^{1/2} =$$

$$\left(\frac{1}{\lambda_i} + \frac{1}{\tilde{\lambda}_{i+1}}\right)\left[(Av,v)+(Aw,w)\right] - \frac{1}{\lambda_i}(Av,v) - \frac{1}{\tilde{\lambda}_{i+1}}(Aw,w) + 2\left[\frac{1}{\lambda_i}(Av,v)\frac{1}{\tilde{\lambda}_{i+1}}(Aw,w)\right]^{1/2} =$$

$$\left(\frac{1}{\lambda_i} + \frac{1}{\tilde{\lambda}_{i+1}}\right)\left[(Av,v)+(Aw,w)\right] - \left\{\left[\frac{1}{\lambda_i}(Av,v)\right]^{1/2} - \left[\frac{1}{\tilde{\lambda}_{i+1}}(Aw,w)\right]^{1/2}\right\}^2 \le$$

$$\left(\frac{1}{\lambda_i} + \frac{1}{\tilde{\lambda}_{i+1}}\right)\left[(Av,v)+(Aw,w)\right].$$

As v and w are A-orthogonal, that is, $(Av,w) = 0$, we have

$$(Av,v)+(Aw,w) = (Av,v) + 2(Av,w) + (Aw,w) = (A(v+w), v+w) = (Au,u)$$

which implies

$$(Bu,u) \le \left(\frac{1}{\lambda_i} + \frac{1}{\tilde{\lambda}_{i+1}}\right)(Au,u)$$

whence

$$\frac{(Bu,u)}{(Au,u)} \le \frac{1}{\lambda_i} + \frac{1}{\tilde{\lambda}_{i+1}}. \tag{3.50}$$

Inequality (3.50) holds for every $u \in \tilde{D}_i$, therefore it holds also for the supremum of $(Bu,u)/(Au,u)$, that is,

$$\sup_{u \in \tilde{D}_i} \frac{(Bu,u)}{(Au,u)} \le \frac{1}{\lambda_i} + \frac{1}{\tilde{\lambda}_{i+1}}$$

which by (3.43) implies

$$\frac{1}{\tilde{\lambda}_i} \le \frac{1}{\lambda_i} + \frac{1}{\tilde{\lambda}_{i+1}}. \tag{3.51}$$

Due to the decomposition (3.39), inequality (3.51) is valid for $i = 1, 2, ..., n - 1$. Taking the notation (3.42) into consideration we obtain the sequence of inequalities

$$\frac{1}{\tilde{\lambda}_0} \le \frac{1}{\lambda_1} + \frac{1}{\tilde{\lambda}_2}, \quad \frac{1}{\tilde{\lambda}_2} \le \frac{1}{\lambda_2} + \frac{1}{\tilde{\lambda}_3}, \quad ... \quad, \frac{1}{\tilde{\lambda}_{n-1}} \le \frac{1}{\lambda_{n-1}} + \frac{1}{\lambda_n}$$

from which the statement of the theorem follows.

R e m a r k s . In the proof it was important that B is positive. Conservative approximation is therefore guaranteed only for problems having positive eigenvalues. In the Theorem it is not required that (Bu,u) is completely continuous with respect to (Au,u) in every subproblem $(u \in D_i, i = 1, 2, ..., n)$. Thus, the Theorem is valid for degenerate subproblems whose spectrum has only one point and it is an eigenvalue with infinite multiplicity, or whose spectrum is continuous. In the case of continuous spectrum, however, we cannot talk about the least eigenvalue but the infimum of the *Rayleigh* quotient defined by (3.43). Such problems can arise for buckling of bars where the effect of shearing forces on the deflection is taken into consideration.

There are problems where by application of the *Föppl* method (partial rigidizing), the conditions of the *Dunkerley* theorem are satisfied, and (3.38) is due to the *Dunkerley* theorem. In such a case the *Föppl-Papkovich* formula is identical to the *Dunkerley* formula. There are, however, cases for which the *Föppl-Papkovich* formula is different from the *Dunkerley* formula, and the *Föppl-Papkovich* formula approximates the exact value with a smaller error than the *Dunkerley* formula. For each of the two sorts of cases we will show an example.

3.2.1 *Example.* Consider the buckling of a bar of constant cross-section with hinged ends, subjected to compressive force N at its ends. The differential equation and the boundary conditions of the problem are:

$$EI_y u'''' + Nu'' = 0, \qquad (3.52)$$

$$u(0) = u''(0) = u(l) = u''(l) = 0. \qquad (3.53)$$

By the notation $Au = EI_y u''''$, $\lambda N_1 = N$, $Bu = N_1 u''$ the eigenvalue problem (3.52), (3.53) can be written in the form $(A - \lambda B)u = 0$. Equation (3.52), however, can be written in the form

$$u'''' + \frac{N}{EI_y} u'' = 0 \qquad (3.54)$$

which with the notation $\overline{A} u = u''''$, $\lambda N_1 = N$, $\overline{B} u = N_1 u'' / (EI_y)$ reads $(\overline{A} - \lambda \overline{B})u = 0$. Subdivide the length of the bar into n parts with points $0 = z_0 < z_1 < z_2 < ... < z_{n-1} < z_n = l$ (Fig. 3.5). The value of the stiffness parameter on every segment is EI_y. Introduce the following notation:

$$K_i = \begin{cases} EI_y, & \text{if} & z_{i-1} < z \le z_i, \\ \infty, & \text{if} & z \le z_{i-1} \quad \text{or} \quad z > z_i, \end{cases}$$

$$\overline{B}_i u = -N_1 \frac{u''}{K_i}, \qquad i = 1, 2, ..., n.$$

It is clear that $\overline{B} = \sum_{i=1}^{n} \overline{B}_i$, and the conditions of the *Dunkerley* theorem (Theorem 3.1) are satisfied. Therefore, if λ_i denotes the least eigenvalue of the eigenvalue problem $(\overline{A} - \lambda \overline{B}_i)u = 0$, then (3.4) holds, but considering the physical background it expresses the inequality (3.38), as by partial rigidizing the bar we have defined A-orthogonal subspaces of displacement functions u. This can be shown as follows. The ith A-orthogonal subspace D_i consists of the four times differentiable functions u satisfying boundary conditions (3.53) such that - because of rigidizing - u is linear outside the interval $z_{i-1} < z \le z_i$. (We note here that for $i = 2, 3, ..., n\text{-}1$ the boundary conditions $u''(0) = u''(l) = 0$ are automatically satisfied.) Let $u \in D_i$, $v \in D_j$ $(i \neq j)$ be arbitrary elements, then by integration by parts

$$(Au, v) = \int_0^l EI_y u''''v \, dz = \left[EI_y u''' v \right]_0^l - \left[EI_y u'' v' \right]_0^l + \int_0^l EI_y u'' v'' \, dz.$$

In this expression the integrated parts vanish in consequence of the boundary conditions. The second derivative of the function u and v vanishes outside the interval $z_{i-1} < z \leq z_i$ and $z_{j-1} < z \leq z_j$, respectively. Since these intervals are disjoint, it follows that $u''v''$ is identically equal to zero. Therefore $(Au, v) = 0$, that is, u and v are A-orthogonal, consequently D_i and D_j are A-orthogonal also. So, the conditions of the *Föppl-Papkovich* theorem (Theorem 3.2) are satisfied, too. By writing the *Rayleigh* quotient we also can show that the least eigenvalues of the eigenvalue problems $\left(\overline{A} - \lambda \overline{B}_i \right) u = 0$, $u \in D_A$ and $\left(A - \lambda B \right) u = 0$, $u \in D_i$ are the same: λ_i. In this example therefore the *Dunkerley* theorem and the *Föppl-Papkovich* theorem result in the same formula.

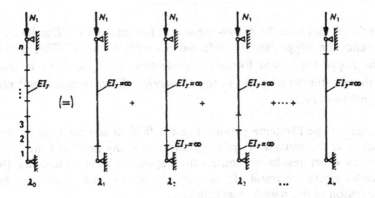

Fig. 3.5 Partial rigidizing of the hinged bar.

3.2.2 *Example.* Consider torsional buckling of a bar of height l with the lower end built in and the upper end free . The equilibrium equation and the boundary conditions with the notation used in Section 3.1 are

$$\begin{bmatrix} A_{11} & 0 \\ 0 & A_{22} \end{bmatrix} \begin{bmatrix} u \\ \varphi \end{bmatrix} - \lambda \begin{bmatrix} B_{11} & B_{12}^{(1)} \\ B_{21}^{(1)} & B_{22}^{(1)} \end{bmatrix} \begin{bmatrix} u \\ \varphi \end{bmatrix} = 0, \tag{3.55}$$

$$u(0) = u''(0) = u'(l) = u'''(l) = 0, \tag{3.56}$$

$$\varphi(0) = \varphi''(0) = \varphi'(l) = \varphi'''(l) = 0. \tag{3.57}$$

Equation (3.55) with notation (3.22) and (3.24) takes the form

$$Aw - \lambda B_1 w = 0. \tag{3.58}$$

Introduce the notation

$$B_1^{(1)} = \begin{bmatrix} B_{11} & 0 \\ 0 & 0 \end{bmatrix}, \quad B_1^{(2)} = \begin{bmatrix} 0 & 0 \\ 0 & B_{22}^{(1)} \end{bmatrix}, \quad B_1^{(3)} = \begin{bmatrix} 0 & B_{12}^{(1)} \\ B_{21}^{(1)} & 0 \end{bmatrix}.$$

It is clear that $B_1 = B_1^{(1)} + B_1^{(2)} + B_1^{(3)}$. Let λ_1, λ_2, λ_3 denote the least eigenvalues of the eigenvalue problems

$$\left(A - \lambda B_1^{(1)}\right)w = 0,$$
$$\left(A - \lambda B_1^{(2)}\right)w = 0,$$
$$\left(A - \lambda B_1^{(3)}\right)w = 0,$$

respectively. If λ_0 denotes the least eigenvalue of the problem (3.58), then due to the *Dunkerley* theorem we obtain:

$$\frac{1}{\lambda_0} \le \frac{1}{\lambda_1} + \frac{1}{\lambda_2} + \frac{1}{\lambda_3}.$$

If EI_y and GI_t are the two stiffness parameters in the *Föppl* method, then the the *Föppl-Papkovich* theorem results in the inequality

$$\frac{1}{\lambda_0} \le \frac{1}{\lambda_1} + \frac{1}{\lambda_2}.$$

It can be seen that in this case the *Föppl-Papkovich* formula and the *Dunkerley* formula are not the same, and the *Föppl-Papkovich* formula is more accurate. (This is obvious, as λ_1 and λ_2 in the *Föppl-Papkovich* formula are identical to λ_1 and λ_2 in the *Dunkerley* formula, but there is the term $1/\lambda_3$ in the *Dunkerley* formula that is not present in the *Föppl-Papkovich* formula.)

3.2.3. *Example.* (The Plantema paradox) In the field of sandwich structures the *Föppl* method is known as the *method of split rigidities.* In many problems the method of split rigidities provides exact results or conservative approximations. Plantema [6], however found an example where the result of this method was *not* on the safe side. Now we will investigate the reason of this paradoxical behaviour.

Consider the buckling of a bar of height H with the lower end built in and the upper end hinged, subjected to a concentrated vertical force N at the hinged end (Fig. 3.6). Suppose that the bar undergoes both bending and shearing deformations. The displacement w of a point of the axis of the bar is composed of two parts: displacement w_B due to the bending deformation and displacement w_S due to the shearing deformation, that is,

$$w = w_B + w_S.$$

The equilibrium of the bar - as is usual in the literature [1] - can be expressed by using w_B only. If B_0 and S denote the bending and the shear rigidities, and $\lambda = N$ the equilibrium equation takes the form

$$w_B'''' - \lambda\left(\frac{w_B''''}{S} - \frac{w_B''}{B_0}\right) = 0 \tag{3.59}$$

and the boundary conditions are

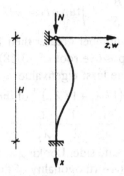

Fig. 3.6 Buckling of a thin-faced sandwich bar with the lower end built in and the upper end hinged

$$w_B(0) = w_B''(0) = w_B(H) - \frac{B_0}{S} w_B''(H) = w_B'(H) = 0. \tag{3.60}$$

The eigenvalue problem (3.59), (3.60) for $S = \infty$ leads to the eigenvalue problem

$$w_B'''' + \lambda \frac{w_B''}{B_0} = 0, \tag{3.61}$$

$$w_B(0) = w_B''(0) = w_B(H) = w_B'(H) = 0, \tag{3.62}$$

and for $B_0 = \infty$ leads to the eigenvalue problem

$$w_B'''' - \lambda \frac{w_B''''}{S} = 0, \tag{3.63}$$

$$w_B(0) = w_B''(0) = w_B''(H) = w_B'(H) = 0. \tag{3.64}$$

Let λ_0, λ_1, λ_2 denote the least eigenvalue of the eigenvalue problem (3.59, 60), (3.61, 62), (3.63, 64), respectively. Introduce the notation

$$\alpha^2 = \frac{N}{B_0\left(1 - \frac{N}{S}\right)}. \tag{3.65}$$

As $N = \lambda$ it is easy to show that the reciprocal of the least eigenvalue of the eigenvalue problem

$$w_B'''' - \lambda \left(\frac{w_B''''}{S} - \frac{w_B''}{B_0}\right) = 0, \tag{3.66}$$

$$w_B(0) = w_B''(0) = w_B(H) = w_B'(H) = 0 \tag{3.67}$$

is equal to $1/\lambda_1 + 1/\lambda_2$. This eigenvalue comes from the least positive solution of the characteristic equation

$$\tan \alpha H = \alpha H. \tag{3.68}$$

The eigenvalue problem (3.59, 60), however, leads to the characteristic equation

$$\left(1+\frac{B_0}{S}\alpha^2\right)\tan\alpha H = \alpha H \tag{3.69}$$

where the coefficient of $\tan\alpha H$ is a number greater then 1. It follows that the least positive root of (3.69) is less than the least positive root of (3.68), and due to (3.68) the less α is, the less N (or λ) is; we obtain that the least eigenvalue λ_0 of the eigenvalue problem (3.59, 60) is less than the least eigenvalue $(1/\lambda_1 +1/\lambda_2)^{-1}$ of the eigenvalue problem (3.66, 67)

$$\frac{1}{\lambda_0} > \frac{1}{\lambda_1}+\frac{1}{\lambda_2},$$

that is, tha *Föppl* method is on the unsafe side. Looking at the problem (3.59, 60), however, the explanation is not at all obvious. (*A*-orthogonality of the bending mode and the shearing mode does not occur, the operators in the problem are not symmetric, and the boundary conditions change with partial rigidizing of the bar.)

To resolve the paradox situation let us describe the problem using both the bending displacement w_B and the shearing displacement w_S explicitly. The equilibrium of the bar in a buckled shape can be written by the system of differential equations

$$B_0 w_B'''' + \lambda(w_B'' + w_S'') = 0, \tag{3.70}$$
$$-S w_S'' + \lambda(w_B'' + w_S'') = 0, \tag{3.71}$$

and the boundary conditions

$$w_B(0) = 0, \qquad w_B''(0) = 0, \qquad w_B(H)-\frac{B_0}{S}w_B''(H) = 0, \qquad w_B'(H) = 0, \tag{3.72}$$
$$w_S(0) = 0, \qquad w_S(H) + w_B(H) = 0. \tag{3.73}$$

By introducing the notation

$$A_{11}w_B = B_0 w_B'''', \tag{3.74}$$
$$A_{22}w_S = -S w_S'', \tag{3.75}$$
$$B_{11}w_B = -w_B'', \tag{3.76}$$
$$B_{22}w_S = -w_S'' \tag{3.77}$$

we obtain the eigenvalue problem

$$\begin{bmatrix} A_{11} & 0 \\ 0 & A_{22} \end{bmatrix}\begin{bmatrix} w_B \\ w_S \end{bmatrix} - \lambda \begin{bmatrix} B_{11} & B_{22} \\ B_{11} & B_{22} \end{bmatrix}\begin{bmatrix} w_B \\ w_S \end{bmatrix} = 0 \tag{3.78}$$

which with the notation

$$A = \begin{bmatrix} A_{11} & 0 \\ 0 & A_{22} \end{bmatrix}, \qquad B = \begin{bmatrix} B_{11} & B_{22} \\ B_{11} & B_{22} \end{bmatrix}, \qquad w = \begin{bmatrix} w_B \\ w_S \end{bmatrix} \tag{3.79}$$

can be written in the form $(A - \lambda B)w = 0$. It can be established that the bending displacement w_B and the shearing displacement w_S as vectors

$$\begin{bmatrix} w_B \\ 0 \end{bmatrix}, \qquad \begin{bmatrix} 0 \\ w_S \end{bmatrix}$$

are *A*-orthogonal. Operator B is symmetric and positive, and (Bw,w) is completely continuous with respect to (Aw,w). *Operator A, however, is not symmetric!* This, however, is not the only reason of the paradoxical behaviour. The question of why the bound goes the wrong way was answered by H.F. Weinberger (private communication): Since $B_0 w_B''' + S w_S'' = 0$ and $w_B''(0) = w_S(0) = 0$, we must have $B_0 w_B'' + S w_S = cx$ for some constant c. Since $w_S(H) = -w_B(H)$, the second boundary condition in (3.73) says that $c = 0$. The Föppl bound appears to come from splitting the space of displacement pairs (w_B, w_S) into the subspaces of pairs $(u,0)$ and $(0,v)$. To satisfy the imposed conditions, one must have $u(0) = u(H) = u'(H) = 0$ and $v(0) = v(H) = 0$. Then any linear combination of such pairs satisfies the additional constraint $w_B(H) = 0$. In other words, the bound given by the method of split rigidities is a lower bound for the critical buckling load of a more constrained problem, but not for the lower buckling load of the original problem. *In mathematical terms, the two subspaces do not span the original space.*

To apply Föppl correctly, one needs to add the supremum on the space of function pairs orthogonal to both of the above subspaces. It is easily seen that this is a one-dimensional space which is spanned by the function pair $(r,-x)$, where

$$r(x) = \frac{1}{2} x \left(3 - \frac{x^2}{H^2} \right).$$

This adds the term

$$\frac{H^2}{5 \left(3 B_0 + H^2 S \right)}$$

to the bound for $1/\lambda$. With this addition, the bound should be correct.

3.2.4 *Example.* Determine the critical load of the pin-ended laced column subjected to concentrated force N at its end (Fig. 3.7a). The column is built up of chord members of length a and cross-sectional area A_0, and of brace members of length d_1 and d_2 and cross-sectional area A_1 and A_2 such that the members are connected to each other by ideal joints. On the basis of the continuum analogy , the critical load of the laced column is calculated similar to that of the compressed bars having shearing deformation (sandwich bars with thin face). The difference between the continuum and the truss only is that for the continuum the shear rigidity is defined for the cross-section of the bar, and the angular deformation (shearing strain) is defined for an infinitesimal length, but for the truss both the shear rigidity and the angular deformation is defined on a finite length a.

Let us suppose first that all the bracing members are infinitely rigid $(EA_1 = EA_2 = \infty)$. Then the column has only bending deformation (due to the elongation of the chord members), and the critical load of the column - with a good approximation - is equal to the *Euler* load:

$$N_E = \frac{\pi^2 E I_0}{H^2} \tag{3.80}$$

where $I_0 = A_0 b^2 / 2$.

Then let us suppose that the chord members are infinitely rigid $(EA_0 = \infty)$, but the bracing members are not. In this case the column can get to a critical state by shearing deformation due to elongation of the bracing members. Let the critical load for shearing deformation be denoted by N_S. According to the *Föppl* summation, the reciprocal of the critical load N_{cr} of the column is

$$\frac{1}{N_{cr}} \approx \frac{1}{N_E} + \frac{1}{N_S}. \tag{3.81}$$

N_S itself can be determined again by the *Föppl* summation. Apart from its ends the column can be considered to be composed of equal rhombic cells braced by diagonal bars. The shearing deformation is due to the elongation of these diagonal bars. Since the column can be composed from rhombi in two different ways (with diagonals with angle of inclination (a) φ_1 and (b) φ_2), the shearing deformation is composed from two independent components. The critical loads associated to these two components can be determined separately.

Consider first the case where all the bars, except the bracing members with inclination of φ_1, are infinitely rigid $\left(EA_0 = \infty, EA_2 = \infty\right)$. Now, in the critical state there is a sway of one of the cells. Lengthening e_1 of the bracing bar in the cell causes a lateral displacement δ_1 (Fig. 3.7b). So, the axis of the deformed column - apart from the deformed cell - has an angle δ_1 / H of inclination to the original axis.

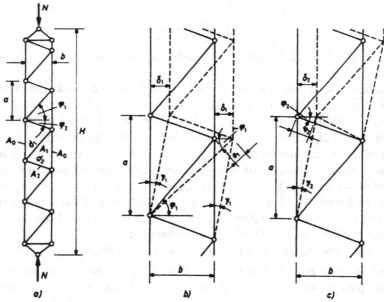

Fig. 3.7 Buckling of a laced column. (a) Layout of the column. Shearing deformation from the elongation of the incline member of length (b) d_1 and (c) d_2

In accordance with Fig. 3.7b

$$\delta_1 = a\gamma_1$$

and

$$e_1 = \delta_1 \cos \varphi_1 = a\gamma_1 \cos \varphi_1.$$

Let us denote the load on this partially rigidized column by N_1. Its critical value can be determined, for instance, by the energy method. The total potential is

$$\Pi = \Pi_{int} + \Pi_{ext}$$

where

$$\Pi_{int} = \frac{1}{2}\frac{EA_1}{d_1}e_1^2 = \frac{1}{2}\frac{EA_1}{d_1}a^2\gamma_1^2 \cos^2 \varphi_1,$$

$$\Pi_{ext} = -N_1 a\frac{\gamma_1^2}{2}.$$

From the condition $\partial\Pi / \partial\gamma_1 = 0$ it follows that

$$N_{1cr} = EA_1 \frac{a}{b}\cos^3 \varphi_1. \tag{3.82}$$

Consider now the case where all the bars, except the bracing members with inclination of φ_2, are infinitely rigid $(EA_0 = \infty, EA_1 = \infty)$. The critical load in this case can be determined from the lengthening of a bracing bar of inclination φ_2 in one of the cells of the other kind (Fig. 3.7c), similar to the previous case. Its value is

$$N_{2cr} = EA_2 \frac{a}{b}\cos^3 \varphi_2. \tag{3.83}$$

Then the critical load of the column having shearing deformations only is obtained by the *Föppl* summation:

$$\frac{1}{N_S} = \frac{1}{N_{1cr}} + \frac{1}{N_{2cr}} \tag{3.84}$$

where N_{1cr} is given by (3.82) and N_{2cr} by (3.83). Knowing (3.80) and (3.84) we obtain the approximate value of the critical load of the laced colum by the expression (3.81).

3.3 The Kollár conjecture

Kollár [7] has published an approximate formula without proof for the determination of the least critical load parameter of compressed continuous bars on elastic supports (Fig. 3.8). Let λ_1 denote the least critical load parameter of the bar for the case where the spring rigidity of the elastic supports is infinitely large, that is, the springs are replaced by rigid supports. Let λ_2 denote the least critical load parameter of the bar for the case where the elastic rigidity of the supports are continuously distributed along the length of the bar, that is, the springs are replaced by an elastic foundation. If the least critical load parameter of the original structure is denoted by λ_0, then Kollár states that:

$$\frac{1}{\lambda_0} \le \frac{1}{\lambda_1} + \frac{1}{\lambda_2}. \tag{3.85}$$

The correctness of formula (3.85) with general validity has not been proved or refuted so far . In this method it is not clear how to distribute the elasticity of the supports, and what kind of distribution function should be considered. Although we cannot prove it, we think

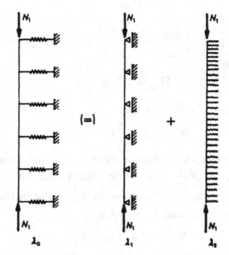

Fig. 3.8 Replacement of the elastic supports

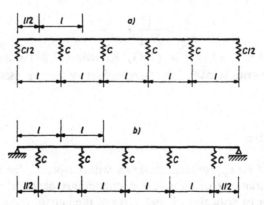

Fig. 3.9 Layout of supports resulting in uniform elastic foundation: (a) with equal spacing,
(b) with equal spring rigidities

that the *Kollár* approximation provides good results for those problems where the distribution of the elasticity of the supports are quite well defined. Such cases are where a bar is supported with equidistant springs of equal rigidity, but where the rigidity of the end springs is the half of the others. In such a case rigidity C of an internal spring should be distributed along the length l, but rigidity $C/2$ of the end spring along the length $l/2$ (Fig. 3.9a). For springs of equal rigidity the arrangement in Fig. 3.9b is suitable. In both cases a uniform elastic foundation is obtained.

We can say something similar about *Vierendel* columns (columns with batten plates) if the equally spaced horizontal beams have the same rigidity, but the uppermost and the bottom beams have the half of their rigidity (Fig. 3.10).

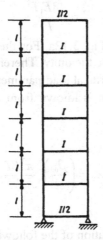

Fig. 3.10 Beam rigidities resulting in uniformly distributed elastic restraint for rotation

In the following the approximation by (3.85) will be illustrated by some examples. Let us define the *Kollár* critical load parameter λ_K by the expression

$$\frac{1}{\lambda_K} = \frac{1}{\lambda_1} + \frac{1}{\lambda_2}. \tag{3.86}$$

3.3.1 *Example.* Let us consider buckling of a continuous beam on $n+1$ supports (Fig. 3.11) in which the middle $n-1$ supports are elastic beams. Let the compressive force be denoted by N, the length of the continuous beam by L and its bending rigidity by EI. Let the

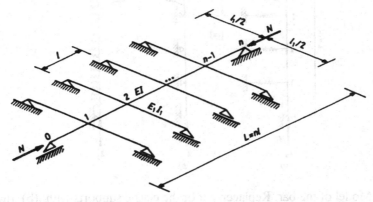

Fig. 3.11 Compressed continuous bar supported by elastic beams

length of the equally spaced cross beams be denoted by l_1 and their bending rigidity by $E_1 I_1$. The cross beams provide elastic supports for the bar with spring rigidity

$$C = \frac{48 E_1 I_1}{l_1^3}.$$

Then we obtain the model shown in Fig. 3.12a. For the sake of simplicity let the reference value N_1 of the compressive force N be the unity. Therefore $\lambda = N$.

The task is to determine the least critical load parameter λ_0 of the bar on elastic supports seen in Fig 3.12a. The exact solution is known from [8]. This critical load parameter is given as follows:

$$\lambda_0 = \left(\frac{\beta_0}{\pi} \right)^2 \frac{\pi^2 E I}{l^2}. \tag{3.87}$$

Here $\beta_0 = \min_{1 \le i \le n-1} \beta_i$ where β_i is the solution of the following equation for fixed i:

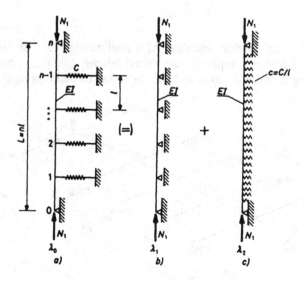

Fig. 3.12 (a) Model of the bar. Replacement of the elastic supports with: (b) rigid supports, (c) elastic foundation

$$\beta^2 \frac{\left(1-\cos\dfrac{i\pi}{n}\right)a-b}{1-\dfrac{b}{1-\cos\dfrac{i\pi}{n}}} - 24\,v = 0 \qquad (i = 1,2,\dots,n-1).$$ (3.88)

Symbols a, b, v in (3.88) are:

$$a = \frac{\beta}{\beta - \sin\beta},$$

$$b = \frac{\beta(1-\cos\beta)}{\beta - \sin\beta},$$

$$v = \frac{E_1 I_1 l^3}{E I l_1^3}.$$

For $n = \infty$ equation (3.88) is reduced to the form

$$\frac{\beta^3(1-\cos\beta)}{\left(\sqrt{\beta}+\sqrt{\sin\beta}\right)^2} - 24\,v = 0.$$ (3.89)

Let us determine now the *Kollár* critical load parameter λ_K of the bar. For that we need the least critical load parameter λ_1 and λ_2 of the bar on rigid supports (Fig. 3.12b) and on elastic foundation (Fig. 3.12c), respectively. λ_1 is easy to determine as it is the *Euler* critical load parameter:

$$\lambda_1 = \frac{\pi^2 EI}{l^2} = \lambda_E.$$ (3.90)

The least critical load parameter of the bar on uniform elastic foundation (with modulus $c = C/l$ of the elastic foundation) can be determined from [1]:

$$\lambda_2 = \min_{m=1,2,\dots} \frac{\pi^2 EI}{l^2}\left[\left(\frac{m}{n}\right)^2 + \left(\frac{n}{m}\right)^2 \frac{48}{\pi^4} v\right].$$ (3.91)

Then due to (3.86), λ_K can be determined by λ_1 and λ_2.

Let $E_1 I_1 = EI$ and $l_1 = 4l$. So, $v = 1/64$. Let us determine the load parameters and the error

$$\Delta = \frac{\lambda_0 - \lambda_K}{\lambda_0}$$

of the *Kollár* approximation for different values of n. We obtained the following results.

$n = 2$
$$\lambda_1 = \lambda_E,$$
$$\lambda_2 = 0.280\ 798\ 0\lambda_E,$$
$$\lambda_K = 0.219\ 236\ 7\lambda_E,$$
$$\lambda_0 = 0.280\ 735\ 4\lambda_E,$$
$$\Delta = 21.9\%.$$

$n = 4$

$$\lambda_1 = \lambda_E,$$
$$\lambda_2 = 0.185\ 691\ 8\lambda_E,$$
$$\lambda_K = 0.156\ 610\ 5\lambda_E,$$
$$\lambda_0 = 0.185\ 535\ 7\lambda_E,$$
$$\Delta = 15.6\%.$$

$n = \infty$

$$\lambda_1 = \lambda_E,$$
$$\lambda_2 = 0.175\ 493\ 4\lambda_E,$$
$$\lambda_K = 0.149\ 293\ 4\lambda_E,$$
$$\lambda_0 = 0.175\ 375\ 8\lambda_E,$$
$$\Delta = 14.9\%.$$

It can be seen that the *Kollár* formula (3.86) is on the safe side in all of the cases.

3.3.2 *Example*. Consider an infinitely long continuous bar on equally spaced equal elastic supports, subjected to constant compressive force (Fig. 3.13). Let the bending rigidity of the bar be EI, the rigidity of the springs be C, the distance between the supports be l. The least critical load parameter of this bar, due to (3.87), is proportional to β_0^2 where β_0 is the solution of the equation (3.89). But now one needs to define

$$\nu = \frac{k}{48}$$

in (3.89), where

$$k = \frac{C}{EI}l^3.$$

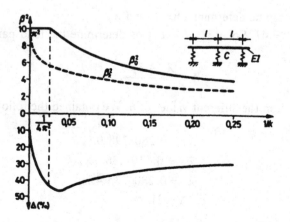

Fig. 3.13 The exact value of the parameter β_0^2 of the critical load, its *Kollár* approximation and the error of the approximation as functions of $1/k$

Let us apply the *Kollár* summation, but instead of the load parameter λ_K, the number β_K^2 corresponding to β_0^2 should be determined, as $\beta_K^2 / \beta_0^2 = \lambda_K / \lambda_0$. With this notation the *Kollár* summation takes the form

$$\frac{1}{\beta_K^2} = \frac{1}{\pi^2} + \frac{1}{2\sqrt{k}}.$$

In Fig. 3.13 there are shown β_0^2 representing the exact solution (solid line), and β_K^2 representing the *Kollár* approximation (dashed line) as functions of $1/k$. It can be ascertained that for every value of $1/k$, the *Kollár* summation is on the safe side. We have also determined the error of the approximation: $\Delta = \left(\beta_0^2 - \beta_K^2\right) / \beta_0^2$, and it is shown in Fig. 3.13 as a function of $1/k$. The maximum error of the *Kollár* conjecture is $\Delta = 45.66\%$ (at $1/k = 0.034$, where $\beta_0 = 3.08$).

3.3.3 *Example.* Consider a two-legged *n*-storey plane frame with rigid restraints at the ends of its columns, each column subjected to a compressive force N (Fig. 3.14). Let the height of the frame be L, the distance between columns be l_1. Let the levels be equally spaced with distance l. Let the bending rigidity of each of the beams of the frame be $E_1 I_1$ and that of the columns of the frame be EI, suppose that the columns are infinitely rigid for axial compression. The beams provide elastic end restraints for the beams with spring rigidity

$$K = \frac{6E_1 I_1}{l_1}.$$

In such a way the half of the frame can be modelled as a compressed bar with elastic restraint at every level (Fig. 3.15a). Let the reference value of the load be the unity.

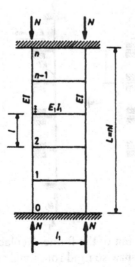

Fig. 3.14 Two-legged plane frame

The task is to determine the least critical load parameter λ_0 of the bar with pointwise rotational elastic restraints shown in Fig. 3.15a. The exact solution can be found in [9]. This critical load parameter can be obtained again from the expression (3.87) with the difference that now β_0 is the minimum of the positive solutions β of the equation

$$\cos\beta + 3\mu\frac{\sin\beta}{\beta} = \cos\frac{2\pi}{n}. \tag{3.92}$$

Here the meaning of parameter μ is:

$$\mu = \frac{E_1 I_1 l}{EI l_1}.$$

Then let us determine the *Kollár* critical load parameter λ_K of the bar. This can be done by the least critical load parameter λ_1 and λ_2 of the sway bar with pointwise rigid restrains for rotation (Fig. 3.15b) and of the bar on elastic foundation for rotation (Fig. 3.15c), respectively. λ_1 is obtained as the *Euler* load of a sway column of length l with built in ends, so it is given by (3.90). The least critical load parameter λ_2 of the bar on elastic foundation for rotation - with modulus $k = K/l$ of continuous rotational elastic support - is the least eigenvalue of the eigenvalue problem

$$EIu'''' - ku'' + \lambda u'' = 0,$$
$$u(0) = u'(0) = u(l) = u'(l) = 0$$

and can be given in the form

$$\lambda_2 = \left[\left(\frac{2}{n}\right)^2 + \frac{6\mu}{\pi^2}\right]\frac{\pi^2 EI}{l^2}. \tag{3.93}$$

Hence, as (3.90) and (3.93) are known, it is easy to obtain λ_K due to (3.86).

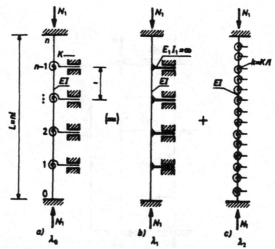

Fig. 3.15 (a) Model of the column of the frame. Replacement of the pointwise elastic rotational support with: (b) pointwise rigid rotational support, (c) continuous elastic restraint for rotation

Let $E_1 I_1 = EI$, $l_1 = l$. So, $\mu = 1$. Let us determine the parameters for different values of n and the error of the *Kollár* approximation. Our calculations have resulted in the following:

$n = 2$

$$\lambda_1 = \lambda_E,$$
$$\lambda_2 = 1.607\ 927\ 1\lambda_E,$$
$$\lambda_K = 0.616\ 553\ 7\lambda_E,$$
$$\lambda_0 = \lambda_E,$$
$$\Delta = 38.3\%.$$

$n = 4$

$$\lambda_1 = \lambda_E,$$
$$\lambda_2 = 0.857\ 927\ 1\lambda_E,$$
$$\lambda_K = 0.461\ 765\ 8\lambda_E,$$
$$\lambda_0 = 0.610\ 985\ 7\lambda_E,$$
$$\Delta = 24.4\%.$$

$n = 20$

$$\lambda_1 = \lambda_E,$$
$$\lambda_2 = 0.617\ 927\ 1\lambda_E,$$
$$\lambda_K = 0.381\ 925\ 2\lambda_E,$$
$$\lambda_0 = 0.404\ 689\ 2\lambda_E,$$
$$\Delta = 5.6\%.$$

$n = \infty$

$$\lambda_1 = \lambda_E,$$
$$\lambda_2 = 0.607\ 927\ 1\lambda_E,$$
$$\lambda_K = 0.378\ 081\ 3\lambda_E,$$
$$\lambda_0 = 0.395\ 809\ 1\lambda_E,$$
$$\Delta = 4.5\%.$$

The *Kollár* formula (3.86) in this example also is conservative for any value of n. It is worth mentioning that, if elastic supports are replaced with elastic foundation only, and the modifying effect of the rigid supports are not taken into account, then in both problems 3.3.1 and 3.3.3 the approximation is not on the safe side ($\lambda_2 > \lambda_0$).

3.3.4 *Example.* Consider an infinitely long continuous bar on equally spaced equal elastic rotational supports, subjected to constant compressive force (Fig. 3.16). Let the bending rigidity of the bar be EI, the rigidity of the rotational (torsional) springs be K, the distance between the springs be l. The least critical load parameter of this bar is proportional to β_0^2 where β_0 is the minimum of the positive solutions of equation (3.92) for $\cos(2\pi/n) = 1$. But now in (3.92)

$$\mu = \frac{Kl}{6EI}$$

should be taken into account. Let us apply the *Kollár* summation, but instead of the load parameter λ_K, the number β_K^2 is determined, similar to the case in 3.3.2. Taking $n = \infty$ into consideration in (3.93), the *Kollár* summation is obtained in the form

$$\frac{1}{\beta_K^2} = \frac{1}{\pi^2} + \frac{1}{\dfrac{Kl}{EI}}.$$

In Fig. 3.16 β_0^2 representing the exact solution is shown as a function of $EI/(Kl)$. The error Δ of the approximation is also shown there. The maximum error of the *Kollár* conjecture is

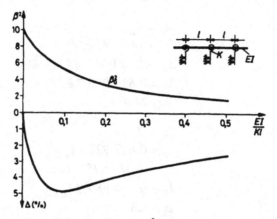

Fig. 3.16 The exact value of the parameter β_0^2 of the critical load and the error of the *Kollár* approximation as functions of $EI/(Kl)$

$\Delta = 4.82\%$ (at $EI/(Kl) = 0.098$). In consequence of a very good agreement between β_0^2 and β_K^2, one cannot distinguish between their graphs in Fig.3.16. It can be ascertained also here that the *Kollár* summation is on the safe side for any value of $EI/(Kl)$.

3.3.5   *Example.* Let the task now be the investigation of torsional buckling of a compressed braced plane bar with open chord sections (Fig. 3.17). The braced bar can have the following buckling modes:
(a) buckling in lateral direction (Fig. 3.18a),
(b) pure torsional buckling (Fig. 3.18b),
(c) twist of the chords with bending of the bracing members (Fig. 3.18c).
The critical load corresponging to the mode (a) and (b) can be determined quite easily. For mode (c), however there is not a known exact solution in a closed form. Therefore an approximatation of the critical load can be composed of parts by the *Kollár* summation. First, the critical load of the chord should be determined for torsional buckling under the assumption that at the nodes the chord is rigidly restrained against twist. This critical load is $P_{cr,rigid}$. Then we determine the critical load of the chords considered as a bar on a

continuous elastic torsional support (elastic restraints provided by the bracing members are continuously distributed along the length of the chords). This critical load is $P_{cr,elastic}$. The critical load $P_{cr,twist}$ for torsional buckling of the chord with elastic torsional restraints at the nodes can be obtained from these two critical loads:

$$\frac{1}{P_{cr,twist}} \approx \frac{1}{P_{cr,rigid}} + \frac{1}{P_{cr,elastic}}.$$

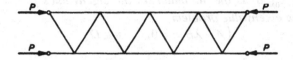

Fig. 3.17 Plane braced bar

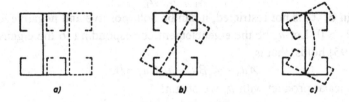

Fig. 3.18 Buckling modes: (a) buckling in lateral direction, (b) pure torsional buckling, (c) twist of the chords with bending of the bracing members

3.4 The Melan theorem

It is known that the natural frequencies of a transversely vibrating bar change under a static axial force. The problem of the free vibration of a bar subjected also to a static axial force is a very old problem. We do not know exactly who was the first to treat this problem. Melan [10] in 1917 published an approximate solution (3.94) for this problem for the case of a concentrated mass at the mid-span of a simply supported bar (but not in this form):

$$\frac{\omega_k^2}{\omega_0^2} = 1 - \frac{N}{N_E} \tag{3.94}$$

where N_E is the *Euler* critical load of the bar, ω_0 is the least natural circular frequency of the bar free from axial force, ω_k is the least natural circular frequency of the bar subjected to axial static force N.   Here N is negative if it is a tensile force. Later it was shown that (3.94) is valid with sign $\leq$, instead of $=$, and equality holds for the case only when the vibration mode and the buckling mode are the same.

In the following we present a general theorem for this problem, and in the applications the *compressive* force will be considered positive.

T h e o r e m . 3.3 *Let A, B, C be symmetric linear operators in a Hilbert space H such that A is positive definite, B and C are positive and $D_A \subset D_B \cap D_C \subset H$. Suppose that (Bu,u) and (Cu,u) is completely continuous with respect to (Au,u), $u \in D_A$. Let λ_0 and μ_0 be a pair of associated eigenvalues of the two-parameter eigenvalue problem*

$$(A - \lambda B - \mu C)u = 0, \qquad u \in D_A \tag{3.95}$$

such that for a fixed λ_0, μ_0 is the minimum of the eigenvalues μ. Let λ_1 and μ_1 be the least eigenvalue of the eigenvalue problem

$$(A - \lambda B)u = 0, \qquad u \in D_A \tag{3.96}$$

and

$$(A - \mu C)u = 0, \qquad u \in D_A, \tag{3.97}$$

respectively. If $\lambda_0 \le \lambda_1$ and $\mu_0 \ge 0$, then

$$\frac{1}{\lambda_0} \le \frac{1}{\lambda_1} + \frac{\mu_0}{\lambda_0}\frac{1}{\mu_1}. \tag{3.98}$$

P r o o f. Sign of λ_0 is not restricted, it can be both positive and negative. Consider first the case where $\lambda_0 > 0$. Let u_0 be the eigenelement corresponding to the eigenvalue pair λ_0 and μ_0. Then (3.95) holds, that is,

$$Au_0 - \lambda_0 Bu_0 - \mu_0 Cu_0 = 0.$$

After forming its scalar product with u_0 we obtain:

$$\frac{1}{\lambda_0} = \frac{(Bu_0,u_0)}{(Au_0,u_0)} + \frac{\mu_0}{\lambda_0}\frac{(Cu_0,u_0)}{(Au_0,u_0)}. \tag{3.99}$$

Now producing the eigenvalues λ_1 and μ_1 with the *Rayleigh* quotients, (3.99) yields:

$$\frac{1}{\lambda_0} = \frac{(Bu_0,u_0)}{(Au_0,u_0)} + \frac{\mu_0}{\lambda_0}\frac{(Cu_0,u_0)}{(Au_0,u_0)} \le \sup_{u \in D_A}\frac{(Bu,u)}{(Au,u)} + \frac{\mu_0}{\lambda_0}\sup_{u \in D_A}\frac{(Cu,u)}{(Au,u)} = \frac{1}{\lambda_1} + \frac{\mu_0}{\lambda_0}\frac{1}{\mu_1},$$

that is identical to the statement (3.98) of the theorem.

Let now $\lambda_0 < 0$. For fixed λ_0, eigenvalue μ_0 of the eigenvalue problem (3.95) can be considered as the least eigenvalue of the eigenvalue problem

$$(A - \lambda_0 B)u - \mu Cu = 0, \qquad u \in D_A,$$

which can be expressed by the *Rayleigh* quotient and so

$$\mu_0 = \inf_{u \in D_A}\frac{(Au,u) - \lambda_0(Bu,u)}{(Cu,u)} \le \frac{(Au,u) - \lambda_0(Bu,u)}{(Cu,u)}.$$

As now $\lambda_0 < 0$, with some algebra we obtain the following inequalities:

$$\mu_0\frac{(Cu,u)}{(Au,u)} \le 1 - \lambda_0\frac{(Bu,u)}{(Au,u)} \le 1 - \lambda_0 \sup_{u \in D_A}\frac{(Bu,u)}{(Au,u)},$$

that is,

$$\mu_0 \frac{(Cu, u)}{(Au, u)} \leq 1 - \frac{\lambda_0}{\lambda_1}, \qquad u \in D_A.$$

This inequality holds for every value of $(Cu, u)/(Au, u)$, $u \in D_A$, consequently for its supremum also:

$$\mu_0 \sup_{u \in D_A} \frac{(Cu, u)}{(Au, u)} = \frac{\mu_0}{\mu_1} \leq 1 - \frac{\lambda_0}{\lambda_1}.$$

Hence

$$1 \geq \frac{\lambda_0}{\lambda_1} + \frac{\mu_0}{\mu_1},$$

and since λ_0 is negative, we obtain

$$\frac{1}{\lambda_0} \leq \frac{1}{\lambda_1} + \frac{\mu_0}{\lambda_0} \frac{1}{\mu_1},$$

that is the statement in the theorem.

R e m a r k . Inequality (3.98) in fact provides a lower bound in a domain, and an upper bound in another domain disjoint from the previous one. The two bounds can be obtained by multiplying (3.98) by λ_0:

$$1 \leq \frac{\lambda_0}{\lambda_1} + \frac{\mu_0}{\mu_1} \qquad (0 \leq \lambda_0 < \lambda_1), \tag{3.100}$$

$$1 \geq \frac{\lambda_0}{\lambda_1} + \frac{\mu_0}{\mu_1} \qquad (\lambda_0 < 0). \tag{3.101}$$

3.4.1 *Example*. Consider the vibration problem of the axially loaded bar mentioned at the beginning of Section 3.4. If EI is the bending rigidity of the cross-section of the bar, m is the mass of the bar on a unit length, N is an axial force on the bar (considered positive if it causes compression), and ω is the circular frequency of the vibration, then the equation of transverse vibration of the bar, after eliminating the time variable, is:

$$\left(EIu''\right)'' + \left(Nu'\right)' - \omega^2 mu = 0. \tag{3.102}$$

Under usual boundary conditions (for instance, simple supports) , equation (3.102) determines a symmetric problem, that is, using notation

$$\lambda N_1 = N,$$

operators A, B, C defined by the relationships

$$Au = \left(EIu''\right)'',$$

$$Bu = -\left(N_1 u'\right)',$$

$$Cu = mu$$

under the given boundary conditions are symmetric, and meet conditions of the Theorem 3.3. If notation $\mu = \omega^2$ is introduced, then statement (3.98) holds, where λ_0 is the given axial force parameter, μ_0 is the square of the least natural circular frequency of the bar for λ_0, λ_1 is the least critical load parameter of the non-vibrating bar ($\mu = 0$), and μ_1 is the square of the least natural circular frequency of the bar free from axial force ($\lambda = 0$). Inequality (3.98), or the pair of inequalities (3.100) and (3.101) are suitable to give bounds of the vibration and stability parameters of the bar.

3.4.2 Bounds for the least natural frequency of a bar subject to given axial force.

3.4.2 Bounds for the least natural frequency of a bar subject to given axial force. From (3.100) and (3.101) it immediately follows that

$$\mu_0 \geq \mu_1\left(1 - \frac{\lambda_0}{\lambda_1}\right), \qquad 0 \leq \lambda_0 < \lambda_1,$$

$$\mu_0 \leq \mu_1\left(1 - \frac{\lambda_0}{\lambda_1}\right), \qquad \lambda_0 < 0.$$

It can be seen that, for the square of the least natural frequency of a bar with axial force parameter λ_0, a lower bound is obtained if the bar is in compression, and an upper bound if a tensile force is in the bar (Fig. 3.19).

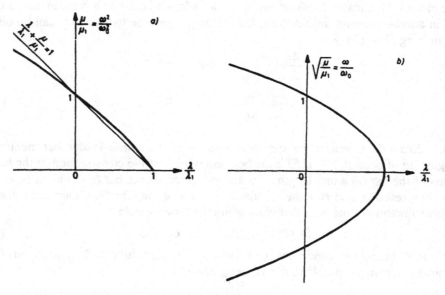

Fig. 3.19 (a) Square of the natural frequency versus the load parameter, and its *Dunkerley* approximation, (b) natural frequency versus load parameter

3.4.3 Bound for the least critical load parameter of a bar.

3.4.3 Bound for the least critical load parameter of a bar. An approximate value of the critical load parameter of the bar free from vibration is sought for. Expressing λ_1 from (3.98) we obtain

$$\lambda_1 \le \frac{\lambda_0}{1 - \dfrac{\mu_0}{\mu_1}}, \tag{3.103}$$

independent of the sign of the force in the bar.

The *Melan* formula (3.98) approximates the least critical load parameter of a structure always on the unsafe side. In most practical cases, however, the error of the approximation (3.103) is not significant. In this way there is a possibility to determine an approximate value of the least critical load parameter of a structure by measuring vibrations. Formula (3.103) is a special case of a more general relationship, which can be obtained in the following way. For two different axial force parameters λ_{01} and λ_{02} we measure the squares μ_{01} and μ_{02} of the least natural circular frequencies of the structure. These values determine two points in the coordinate system λ, μ, and the two points determine a straight line. Points of intersections of this straight line and the coordinate axes provide approximate values of λ_1 and μ_1:

$$\lambda_1 \le \lambda_{01} + \frac{\lambda_{02} - \lambda_{01}}{1 - \dfrac{\mu_{02}}{\mu_{01}}}, \tag{3.104}$$

$$\mu_1 \le \mu_{01} + \frac{\mu_{01} - \mu_{02}}{1 - \dfrac{\lambda_{02}}{\lambda_{01}}}, \qquad \lambda_{01}\lambda_{02} \ge 0, \tag{3.105}$$

$$\mu_1 \ge \mu_{01} + \frac{\mu_{01} - \mu_{02}}{1 - \dfrac{\lambda_{02}}{\lambda_{01}}}, \qquad \lambda_{01}\lambda_{02} \le 0. \tag{3.106}$$

By means of formula (3.104), the least critical load (and from it the bending rigidity) of structures can be determined with good approximation, which actually cannot be subjected to compressive forces. Such structures are, for instance, cables. Interestingly, in such a way, a characteristic compression parameter (the least critical load parameter) of a structure can be determined by a tensional experiment.

3.4.4 *Stability domain.* Stability domain is a set of points in the load parameter space where the structure is in a pre-critical, stable state of equilibrium. Looking at the stability domain of a transversely vibrating bar subjected to an axial force (Fig. 3.19) , it is easy to discover a close analogy with the stability domain in Fig. 1.1b. In the *Dunkerley* theorem only the least positive critical load parameters are taken into account. If the load is considered as a multy-parameter system of forces, then well-posed problems can be defined where some of the parameters are negative. For example, for a two-parameter system the buckling problem makes sense if it is considered not only on the positive quarter of the parameter plane but on the remaining three quarters also. In this way the complete stability domain can be produced, and bounds can be given for the different segments of the boundary curve (boundary hypersurface) of the stability domain, although if some of the parameters are negative the *Dunkerley* theorem is not valid. There exist several different

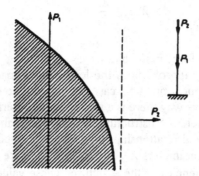

Fig. 3.20 Interaction between two forces in buckling; stability domain

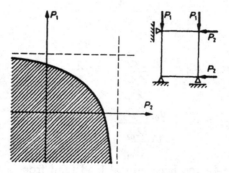

Fig. 3.21 Interaction between two forces in buckling; stability domain

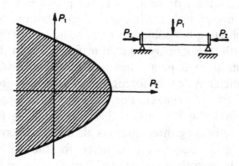

Fig. 3.22 Interaction between forces causing lateral buckling and Euler buckling; stability domain

stability domains for different buckling problems. Some of them are shown in Figs. 3.20, 3.21, 3.22, 3.23. Figs. 3.20 and 3.21 are characteristic of problems where loss of stability occurs, for every subsystem, only for positive load parameter. Figs. 3.22 and 3.23 are characteristic of problems where loss of stability can occur both positive and negative values of the load parameters. Lateral buckling is such a problem, for instance.

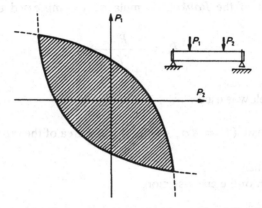

Fig. 3.23 Interaction between two forces causing lateral buckling; stability domain

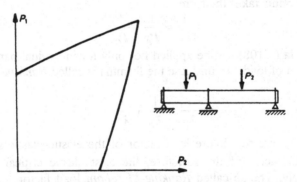

Fig. 3.24 Interaction between two forces causing lateral buckling

Although complete stability domains are used in practice very rarely, it is important to know them, since they provide insight into some unusual interaction relationships like that in Fig. 3.24, for lateral buckling of a continuous beam on three supports, subjected to transverse forces at the middle of the spans. Fig. 3.24 is in fact similar to a part of Fig. 3.23, and the unusual knee-point on the boundary is explained by the fact that the lens-like stability domain in Fig. 3.23 is an intersection of two domains with smooth boundary. Intersection of the two boundaries produces the vertex of the interaction curve.

3.5 The Rankine formula

The *Rankine* formula in fact does not belong to the *Dunkerley* type formulae because it does not relate to elastic structures but elastic-plastic structures. The reason, why the *Rankine* formula is mentioned here, is that its algebraic form in special cases is similar to that of the *Dunkerley* formula.

The original form of the *Rankine* formula of a compressed elastic-plastic bar is the following:

$$P_F \approx \frac{P_P}{1 + a\left(\dfrac{l}{r}\right)^2},$$
(3.107)

where the notation below is used
P_F failure load,
P_P plastic collapse load ($P_P = A\sigma_y$ where A is the area of the cross-section and σ_y is the yield stress),
l the length of the bar,
r radius of gyration of the cross-section,
a a constant.
If the bar is pin-ended, and constant a is selected as

$$a = \frac{\sigma_y}{\pi^2 E},$$

then the *Rankine* formula takes the form

$$\frac{1}{P_F} \approx \frac{1}{P_P} + \frac{1}{P_C}.$$
(3.108)

The *Rankine* formula (3.108) can be applied not only for individual bars but bar structures subjected to a system of loads. In this case the formula is called *Rankine-Merchant* formula, and it has the form

$$\frac{1}{\lambda_F} \approx \frac{1}{\lambda_P} + \frac{1}{\lambda_C}$$
(3.109)

where λ_F, λ_P, λ_C are the failure load factor of the elastic-plastic structure, the rigid-plastic collapse load factor of the structure, the least elastic critical load factor of the structure, respectively. The so-called *Rankine-Merchant* load factor λ_R is defined by the relationship

$$\frac{1}{\lambda_R} = \frac{1}{\lambda_P} + \frac{1}{\lambda_C}.$$

The *Rankine-Merchant* load factor λ_R for certain problems gives a lower bound of the true failure load factor λ_F, but for other problems an upper bound. The *Rankine-Merchant* load factor is in fact an empirical result, therefore it is not possible to stand useful conditions under which the *Rankine-Merchant* load factor is a lower bound of the failure load factor with absolute certainty [11].

Experiments show that for certain structures, the *Rankine-Merchant* formula (3.109) is too conservative. Thus, it was reasonable to fit the formula to the experimental results. This intention has led to the non-linear interaction formulae, many of them have been applied in different design codes. One of them is similar to the *Strigl* formula (3.33):

$$\frac{\lambda_F}{\lambda_P} + \frac{\lambda_F}{\lambda_C} - \frac{\lambda_F^2}{\lambda_P \lambda_C} \approx 1.$$

But another one is also well-known:

$$\left(\frac{\lambda_F}{\lambda_P}\right)^m + \left(\frac{\lambda_F}{\lambda_C}\right)^n \approx 1, \quad m \geq 1, \quad n \geq 1.$$

All these formulae are also empirical; to our knowledge there are no conditions under which these formulae provide lower bounds of the true failure load.

4. CONCLUSIONS

From point of view of practical structural design, we summarise the statements of this chapter as follows.

(1) The method based on the subdivision of rigidity into parts, due to the *Southwell* theorem, gives a conservative approximation of the least elastic critical load parameter only if the buckling problem cannot be defined for the opposite direction of the load; that is, the problem has positive eigenvalues only.

(2) The method based on the subdivision of the load system into parts, due to the *Dunkerley* theorem, in the case of all the structures occurring in practice, gives a conservative approximation of the least positive elastic critical load parameter of the structure.

(3) The method based on partial rigidizing of a structure, due to the *Föppl-Papkovich* theorem, gives a conservative approximation of the least elastic critical load parameter only if the deformations ocurring at different partial rigidizing of the structure are independent, and the buckling problem cannot be defined for the opposite direction of the load; that is, the problem has positive eigenvalues only.

(4) The *Kollár* method based on a transformation of the elastic properties of the investigated structure, according to experiences so far, gives a conservative approximation of the least elastic critical load parameter of the structure. Correctness of this method, however, has not been proved yet. A proof of the *Kollár* conjecture is still needed.

(5) For a transversely vibrating structure subjected to a static axial force, in every case, the *Melan* formula gives a non-conservative estimate of the least elastic critical load parameter of the structure.

(6) Linear and non-linear variants of the *Rankine* formula can give both conservative and non-conservative approximations of the failure load parameter of an elastic-plastic structure. The *Rankine* formula in fact is an empirical relationship which, as an inequality, theoretically cannot be proved with general validity.

(7) The summation theorems and formulae are valid only for bifurcation critical loads in the linear theory of stability. Critical loads representing limit points cannot be included.

(8) The summation theorems are valid only for conservative forces. This follows from the fact that the equation $(A - \lambda B)u = 0$ is in fact an equilibrium equation expressed by displacement u. Forming the scalar product of this equation with u we obtain

$$(Au, u) - \lambda(Bu, u) = 0$$

that yields the *Rayleigh* quotient, if λ is expressed from it. Expression $(Au, u) - \lambda(Bu, u)$ is twice the total potential energy of the system. Thus, it is tacitly supposed that there exists a potential, that is, the forces are conservative.

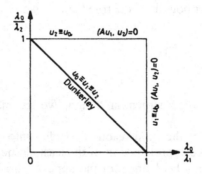

Fig. 4.1 Effect of buckling forms upon the exactness of the *Dunkerley* approximation

(9) An important term in the theory of elastic stability is orthogonality with respect to a symmetric linear operator A, that is, A-orthogonality. [u, $v \in D_A$ are A-orthogonal, if $(Au, v) = 0$ (Section 3.2).] It is easy to show that eigenfunctions (buckling forms) corresponding to different eigenvalues of the problem $(A - \lambda B) = 0$ are A-orthogonal. The approximation due to the *Southwell* and *Dunkerley* theorems is the more exact, the closer the buckling forms of the subproblems are to each other. The formulae are exact, if the buckling forms of the subproblems are identical. The error of the approximation is a maximum if the buckling forms of the subproblems are A-orthogonal. In this case the buckling modes corresponding to the subproblems are not combined in the original problem. In connection with the *Dunkerley* theorem this fact is illustrated in Fig. 4.1 where u_0, u_1, u_2 are buckling shape functions (eigenfunctions) corresponding to critical load parameters (eigenvalues) λ_0, λ_1, λ_2.

Acknowledgement. This work was partially supported by OTKA I/3 Grant No.684 awarded by the Hungarian Scientific Research Foundation.

REFERENCES

1. Timoshenko, S.P. and J.M. Gere: Theory of Elastic Stability, McGraw-Hill, New York, 1961.
2. Weinberger, H.: Variational Methods for Eigenvalue Approximation. SIAM Philadelphia, Pa. 1974.
3. Weinberger, H.: Some mathematical aspects of buckling. Chapter 1 in this volume.
4. Kollár L.(Editor): Special problems of engineering theory of stability (in Hungarian), Akadémiai Kiadó, Budapest, 1991.
5. Strigl, G.: Das nicht lineare Überlagerungsgesetz für die Lösungen von zusammengesetzten Stabilitätsproblemen mit Verzweigungspunkt, Der Stahlbau, 24 (1955), 33-39, 51-61.
6. Plantema, F.J.:Theory and Experiments on the Elastic Overall Instability of Flat Sandwich Plates, Doctoral Thesis, Delft, 1952.
7. Kollár L.: Recent results in the theory of stability through the eyes of a designer (in Hungarian), Magyar Építőipar, 20 (1971), 333-337.
8. Bleich, F.: Buckling Strength of Metal Structures, McGraw-Hill, New York 1952.
9. Kármán, T. and M.A. Biot: Mathematical Methods in Engineering, McGraw-Hill, New York, 1940.
10. Melan, H.: Kritische Drehzahlen von Wellen mit Langsbelastung . Zeitschrift der Österr. Ingenieur- und Architekten-Vereines, 69 (1917), 610-612, 619-621.
11. Horne, M.R.: The Rankine-Merchant load and its application. Chapter 3 in this volume.

PRACTICAL EXAMPLES

L. Kollár
Technical University of Budapest, Budapest, Hungary

ABSTRACT

A great variety of practical examples are presented which demonstrate
the wide applicability of the summation theorems to problems of structural stability. After three simple examples several more complex
problems are discussed involving columns, plates, frames, arches, in
which these theorems can be usefully applied. Finally, practical
problems associated with the stiffening of buildings and the lateral
stability of beams are treated.

INTRODUCTION

The summation theorems of Southwell, Dunkerley and Föppl-Papkovich
have long been available but infrequently used by engineers in routine
design, unless formally included in design codes. This tends to lead
to their application in a stereotyped fashion and masks their more
general usefulness. It is the objective here to demonstrate imaginati-
ve ways of using the summation theorems to solve problems of a less
routine nature.

All summation theorems yield critical loads wich are inferior than (or
equal to) the exact critical loads. However, for simplicity, we shall
use the equality sign in this chapter.

Let us begin with some very simple cases for the three
summation theorems.

1. A simple example for Southwell's Theorem

Let the I-shaped column of variable cross section shown in
Fig. 1-1a be subjected to its own weight p $[N/m]$, considered
constant along the height. We seek estimates for the
critical value of the load causing buckling in the plane of
the drawing.

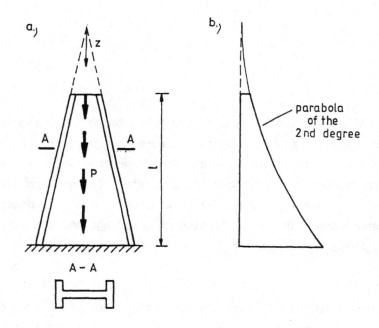

Fig. 1-1 Column with variable cross section subjected to a
uniformly distributed load

If we neglect the web when calculating the moment of
inertia (second moment of area) of the cross section, we
find that the moment of inertia varies with the second power
of the vertical co-ordinate z (Fig. 1-1b). In the literature
we cannot find a ready-to-use formula for this case.

However, we can find formulas for a column under uniformly distributed load with a moment of inertia varying according to the law $I_n * \left(\frac{z}{l}\right)^n$ in [Timoshenko and Gere, 1961]. Hence we decompose the diagram of the moment of inertia into a constant, a linear, and a quadratic component, see Fig. 1-2.

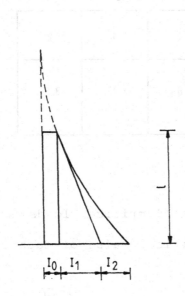

I_0 I_1 I_2

Fig. 1-2 Decomposing the diagram of the moment of inertia

From [Timoshenko and Gere, 1961] we can take the critical load values for every case, see Fig. 1-3.
Using Southwell's Theorem, we obtain a lower bound for the critical load:

$$P_{cr} = \frac{E}{l^2}\left(7.84\,I_0 + 5.78\,I_1 + 3.67\,I_2\right)$$

2. A simple example for Dunkerley's Theorem

A column with constant cross section is acted upon by three concentrated forces of equal magnitude (Fig. 2-1). Let us determine the critical value of these forces.

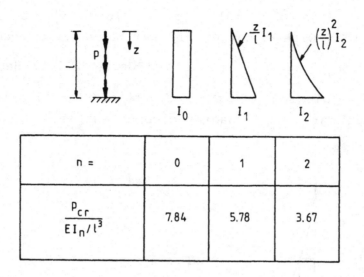

n =	0	1	2
$\dfrac{P_{cr}}{EI_n/l^3}$	7.84	5.78	3.67

Fig. 1-3 Partial critical loads

We first determine the critical value of each force acting separately:

$$P_{1cr} = \frac{\pi^2 EI}{4\, l^2}, \quad P_{2cr} = \frac{\pi^2 EI}{4\left(\frac{2}{3}l\right)^2}, \quad P_{3cr} = \frac{\pi^2 EI}{4\left(\frac{1}{3}l\right)^2},$$

so that, applying Dunkerley's theorem, we have

$$\frac{P}{P_{1cr}} + \frac{P}{P_{2cr}} + \frac{P}{P_{3cr}} = 1,$$

from which we obtain the lower bound of the critical value as:

$$P = P_{cr} = 1.587 \frac{EI}{l^2}.$$

It is also possible to decompose the load in another way. However, we always have to check whether we do not commit an error which reduces safety. That is, if we replace the diagram of the normal forces by a constant and a linearly

varying part as shown in Fig. 2-2, we actually substituted
the vertically hatched areas for the horizontally hatched
ones, i.e. we assumed that these forces act at a lower
height than in reality. Consequently, we commit an error
which loads to an unsafe result. In fact, we thus obtain the
following:

Since for a linearly varying axial force diagram the
critical value of the axial force at the bottom is

$$P_{cr}^{lin} = 7.84\frac{EI}{l^2},$$

so Dunkerley's theorem takes the shape

$$\frac{P/2}{\dfrac{\pi^2 EI}{4\, l^2}} + \frac{3P}{7.84\dfrac{EI}{l^2}} = 1,$$

from which

$$P \doteq P_{cr} = 1.708\frac{EI}{l^2},$$

i.e. 7.6 % higher than before.

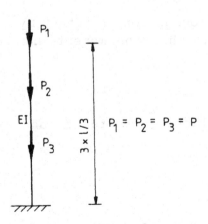

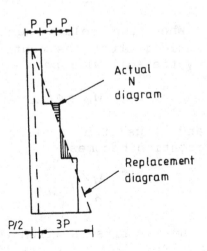

Fig. 2-1 Column loaded by
three concentrated forces

Fig.2-2 Replacement
axial force diagram

3. A simple example for the Föppl-Papkovich Theorem.

Let us consider a column on elastic foundation, loaded by its own weight (Fig. 3-1a). The elastic restraint is exerted by the soil on which the foundation plate of the column rests. The soil has a foundation coefficient c $[N/m^3]$, referred to the soil pressure N/m^2.

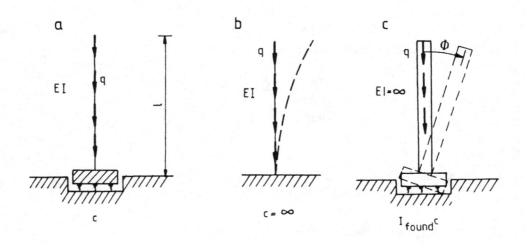

Fig. 3-1   Column on elastic foundation

When the column buckles, its foundation rotates by an angle ϕ about its central point, so that the moment exerted by the soil will be

$$M_{soil} = \phi I_{found} \cdot c,\qquad\qquad (3\text{-}1a)$$

and thus the coefficient of elastic restraint against rotation becomes

$$\frac{M}{\phi} = I_{found} \cdot c.\qquad\qquad (3\text{-}1b)$$

Let us first stiffen the foundation against rotation $(c \rightarrow \infty)$, see Fig. 3-1b. The critical load of the column then becomes

$$\left(ql\right)_{cr}^{bend} = \frac{7.84\,EI}{l^2}. \qquad\qquad (3\text{-}2)$$

Let us now stiffen the column against bending deformation ($EI\rightarrow\infty$), as shown in Fig. 3-1c. The critical load of the rigid column on elastic foundation can be easily calculated as:

$$ql * \phi\frac{l}{2} = \phi\,I_{found}\,c,$$

or $\qquad \left(ql\right)_{cr}^{rot} = \frac{2}{l}I_{found}c. \qquad\qquad (3\text{-}3)$

Using the Föppl-Papkovich theorem the critical load of the original system can be computed:

$$\frac{1}{\left(ql\right)_{cr}} = \frac{1}{\left(ql\right)_{cr}^{bend}} + \frac{1}{\left(ql\right)_{cr}^{rot}}. \qquad\qquad (3\text{-}4)$$

4. Critical load of a column undergoing bending and shear deformation

The critical load of a bar with pure bending deformation will be reduced if the bar undergoes also shear deformation. This phenomenon can be simply treated by the Föppl-Papkovich theorem.

Let us stiffen the bar shown in Fig. 4-1a first against shear deformation. In this way we obtain the well-known Euler formula for the critical load:

$$P_{cr}^{bend} = \frac{\pi^2 EI}{4\,l^2}. \qquad\qquad (4\text{-}1)$$

Next we assume that the bar undergoes shear deformation only. In the deformed bar (Fig. 4-1b) at every cross section a shearing force T arises which can be computed from the loading force P as follows:

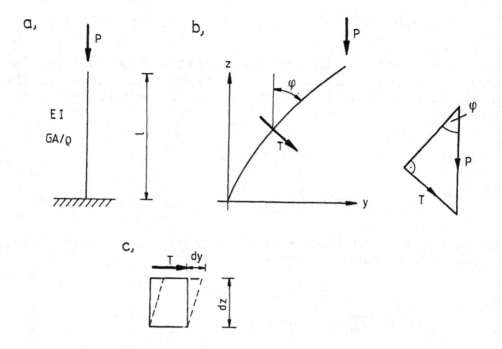

Fig 4-1   Column undergoing bending and shearing deformation

$$T = P \sin \varphi \approx P \tan \varphi = P \frac{dy}{dz}. \qquad (4\text{-}2a)$$

The shearing deformation of a section of length dz of the bar is (Fig. 4-1c):

$$\frac{dy}{dz} = \frac{T}{GA/\rho} \qquad (4\text{-}2b)$$

with ρ as the shape factor. Introducing (4-2a) into (4-2b) and simplifying by dy/dz we obtain

$$P = \frac{GA}{\rho} = P_{cr}^{shear}. \qquad (4\text{-}3)$$

which gives the load value ensuring equilibrium in the deformed state, i.e. the critical load of a column deforming in shear only.

The most striking aspect of this result is that the buckled shape of the column is not defined, it can be an arbitrary line, as contrasted with buckling by bending where the shape of the buckling mode is well defined.

Proceeding now according to the Föppl-Papkovich theorem we obtain for the critical load of a column deforming in bending and shear

$$\frac{1}{P_{cr}} = \frac{1}{P_{cr}^{bend}} + \frac{1}{P_{cr}^{shear}}, \tag{4-4}$$

from which

$$P_{cr} = \frac{P_{cr}^{bend}}{1 + \dfrac{P_{cr}^{bend}}{P_{cr}^{shear}}}. \tag{4-5}$$

This formula is identical to that derived in [Timoshenko and Gere, 1961], i.e. in this case the Föppl-Papkovich theorem yields the exact result. The explanation is given by the theorem of Strigl: the results of the summation theorems are the better, the closer the buckling modes of the partial solutions lie to each other. Since, in this case, the buckling mode of the column with shearing deformation is arbitrary, it can take the shape of that pertaining to buckling by bending, which ensures an exact result.

Let us now investigate the same column loaded by uniformly distributed vertical load p (Fig. 4-2a). The critical load for buckling by bending is

$$p_{cr} = 7.84 \frac{EI}{l^3}. \tag{4-6}$$

The differential equation for buckling by shear is given by (4-2b). In this case the axial force, P, varies according to the law

$$P = p(l - z).$$

The shearing force, T, is given by (4-2a) and takes the particular form

$$T = p(l - z)\frac{dy}{dz}. \tag{4-7}$$

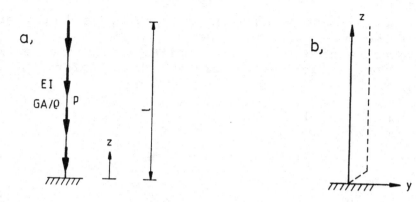

Fig. 4-2   Buckling by shear of a column loaded by
                        distributed forces

Introducing (4-7) into (4-2b) yields

$$\frac{dy}{dz} = \frac{\rho}{GA} p(l-z) \frac{dy}{dz},$$ (4-8a)

or $$\frac{dy}{dz}\left[\frac{GA}{\rho} - p(l-z)\right] = 0.$$ (4-8b)

Eq. (4-8b) is satisfied along the bar only if dy/dz=0, but
at z=0 the expression between square brackets can also be
zero, so that dy/dz can here be arbitrary: the bar buckles
at this point. Since the compressive force assumes here its
maximum value, it is quite obvious that the bar buckles
here. Hence the critical load is

$$(pl)_{cr} = \frac{GA}{\rho}.$$ (4-9)

It can be seen that the buckling mode has a peculiar
shape: at $z=l$ there is a "kink" (which can have any value),
but the whole bar remains straight and vertical (Fig. 4-2b).
This also means that the buckling mode has a well-defined
shape, as contrasted with the previous case where it was
completely arbitrary. Thus if we again use the Föppl-
Papkovich theorem, see (4-4) and (4-5), we obtain a critical
load for buckling by bending and shear, which is lower than

the exact value, since the buckling modes of the two partial
solutions have different shapes.

5. Buckling of a plate supported at points.

Let us consider a plate supported at points arranged
according to a regular pattern (Fig. 5-1), and subjected to
hydrostatic compression $\left(n_x = n_y = n\right)$. We seek the critical value
of n.

The support points allow three (geometrically possible)
buckling modes (Fig. 5-2): two cylindrical ones and a
doubly-curved one represented by contour lines.

The critical compressive forces pertaining to these
buckling modes are:

a)
$$n_{x.cr} = \frac{\pi^2 E t^3}{12\left(1 - v^2\right) a^2}, \qquad n_y \leq n_{x.cr} \qquad\qquad (5\text{-}1)$$

b)
$$n_{y.cr} = \frac{\pi^2 E t^3}{12\left(1 - v^2\right) a^2}, \qquad n_x \leq n_{y.cr} \qquad\qquad (5\text{-}2)$$

with n as the compressive force per unit width, t as the
thickness of the plate and v as Poisson's ratio.

c) The critical compressive force in this case is
availeble in [Timoshenko and Gere, 1961] as

$$n_{cr} = 2 \frac{\pi^2 E t^3}{12\left(1 - v^2\right)\left(\sqrt{2}\, a\right)^2} = \frac{\pi^2 E t^3}{12\left(1 - v^2\right) a^2}. \qquad\qquad (5\text{-}3)$$

Now we have to examine: how we can obtain the critical
compressive force for the plate using the solutions of these
three cases.

In principle we could apply the Föppl-Papkovich theorem,
considering that the plate is in every case stiffened
against the deformations of the two other modes. Since all
three critical compressive forces are, equal, we would
obtain one third of these values as the "combined" critical
load.

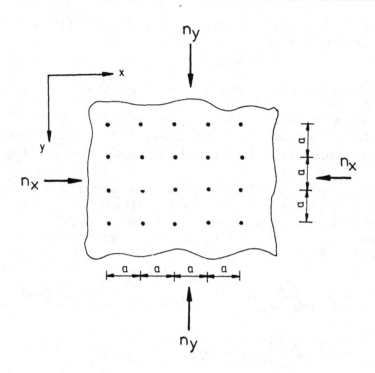

Fig. 5-1   Plate supported at points arranged according to a
rectangular pattern

However, we have to consider that the buckling modes of a
structure are orthogonal to each other, and so are modes a),
b) and c).

Now the theorem of Strigl states that orthogonal modes do
not interfere. Consequently, the critical load is given by
(5-1), (5-2) or (5-3).

All this can be represented by the diagram of Fig. 5-3.
The three critical loads are represented by lines a) and b),
and by point c). As can be seen, as long as $n_x \leq n_y = n_{y,cr}$ the
magnitude of n_x does not influence buckling. The same is true
for n_y if $n_y \leq n_{x,cr}$. Finally, if $n_x = n_y$, then any of the three
buckling modes may develop, but all three modes have the
same n_{cr} value.

The dashed line in Fig. 5-3 corresponds to combinations n_x,
n_y, where $n_x \neq n_y$, see [Timoshenko and Gere, 1961].

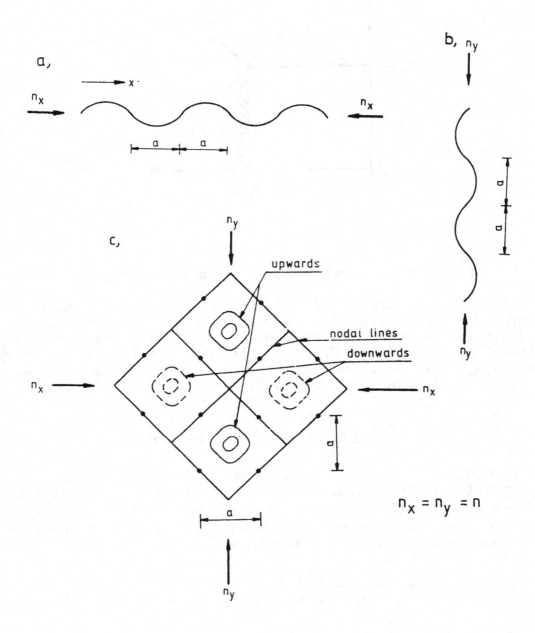

Fig. 5-2 Possible buckling modes of the plate of Fig. 5-1

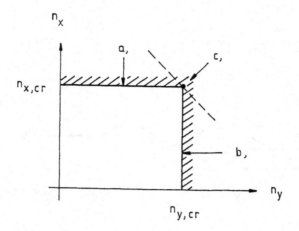

Fig. 5-3   Critical loads of the plate of Fig. 5-1

For other patterns of the supporting points a similar approach can be adopted. E.g. for the plate shown in Fig. 5-4 there exist three cylindrical and one doubly-curved buckling modes, see Fig. 5-5, each of which buckles at the same n_{cr}. For the cylindrical modes a), b) and c) a uniaxial critical compression of the magnitude

$$n_{cr} = \frac{\pi^2 E t^3}{12(1 - v^2)(0.866a)^2} = 1.33\frac{\pi^2 E t^3}{12(1 - v^2)a^2}$$

is valid. The compression perpendicular to the waves does not influence buckling, so that the hydrostatic compression $n_x = n_y = n_{cr}$ is equally likely to cause any of these buckling modes. The doubly-curved buckling pattern d) occurs under a uniform compression of the magnitude

$$n_{cr} = 4\frac{\pi^2 E t^3}{12(1 - v^2)(0.866 \times 2 a)^2} = 1.33\frac{\pi^2 E t^3}{12(1 - v^2)a^2},$$

see [Timoshenko and Gere, 1961], so that also in this case there is only one critical uniform compression which may produce either of the four buckling modes, which do not interact. Thus for this problem with $n_x = n_y$ all four buckling

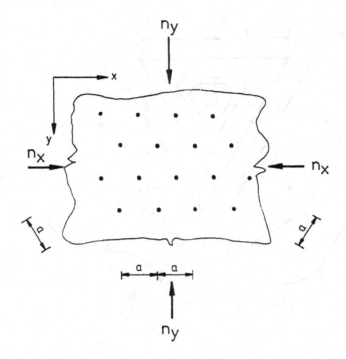

Fig. 5-4 Plate supported at points arranged according
to a triangular pattern

modes are possible at the same critical load factor. The
actual mode experienced will depend on the initial
imperfection of the plate.

6. Buckling of frames

The buckling of a multi-storey, single-bay frame (Fig. 6-
1a) can be analyzed in a very simple way as follows. In the
first step we "smear out" the beams over the height, and in
the second step we take the difference between uniformly
distributed and concentrated beams into consideration.
During buckling the beams deform antisymmetrically. They
exert a rotational stiffness at each connection point given
by $6EI_b/l$, which is to be distributed along the storey
height, h, so that the distributed elastic restraint against
rotation is

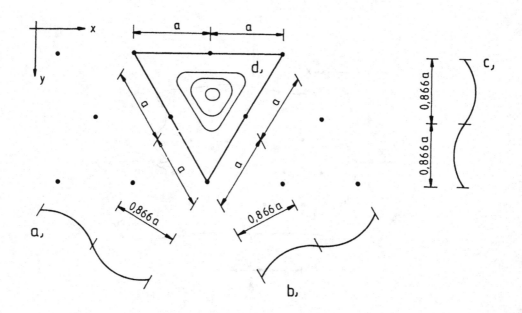

Fig. 5-5 Buckling modes of the plate of Fig. 5-4

$$k = \frac{6\,EI_b}{l\,h}. \tag{6-1}$$

We thus have to solve the buckling problem of the column of Fig. 6-1b, representing half the frame [Csonka, 1961].

The distributed moments m exerted by the elastic restraint are given by

$$m = -k\frac{dy}{dx}. \tag{6-2}$$

We can write the bending moment at a certain cross section n caused by the external load as

$$M = -EI_c\frac{d^2y}{dx^2}, \tag{6-3}$$

and the elastic restraint as

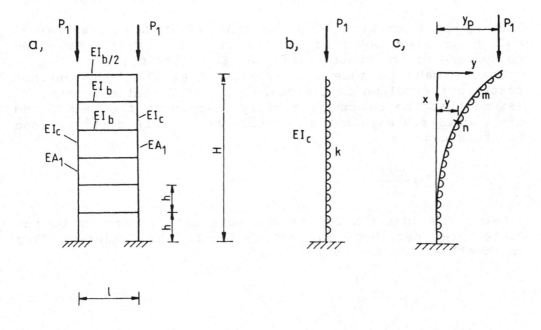

Fig. 6-1 Multi-storey, single-bay frame

$$M = -P_1(y_p - y) + \int_0^x m\, dx, \qquad (6-4)$$

see Fig. 6-1c.

Introducing (6-2) and (6-3) into (6-4), and differentiating twice, we obtain

$$(P_1 - k)\frac{d^2 y}{dx^2} + EI_c \frac{d^4 y}{dx^4} = 0. \qquad (6-5)$$

This equation is identical to that of a compressed bar without elastic restraint, except that the external load P_1 is replacad by $(P_1 - k)$. Thus the eigenvalue $P_{0,cr}$ of the differential equation (for any boundary condition) is equal to $(P_1 - k)$, and the actual (global) critical load is

$$P_{1,cr}^{glob} = P_{0,cr} + k. \qquad (6-6)$$

The (global) critical load of the elastically restrained column is thus obtained as the sum of that of the free column and of the constant of elastic restraint.

Now we have to take into account that the beams do not restrain the column continuously, but at isolated points. We assume that the column is rigidly clamped at the beams, and can buckle sideways between them. The (local) critical load for this case is

$$P_{1,cr}^{loc} = \frac{\pi^2 EI_c}{h^2}$$ (6-7)

Using the addition of inverts we obtain according to the conjecture descibed in Sect.3.3 of the chapter "The Southwell and Dunkerley Theorems":

$$\frac{1}{P_{1,cr}} = \frac{1}{P_{1,cr}^{glob}} + \frac{1}{P_{1,cr}^{loc}}$$ (6-8)

In order to check this result we use the exact solution of the problem to be found in [Kármán and Biot, 1940] for a frame rigidly and laterally immovably clamped at both ends.

Let us assume $h = H/6$, $EI_b h/(EI_c l) = 1$. Introducing the notation

$$\frac{\pi^2 EI_c}{H^2} = P_{cr}^{Euler}$$ (6-9)

we thus obtain

$$P_{1,cr} = 18 P_{cr}^{Euler}.$$

Eq. (6-1) yields

$$k = \frac{EI_b}{EI_c} \frac{h}{l} \frac{6EI_c}{h^2} = 22 P_{cr}^{Euler}$$

Eq. (6-6): $P_{1,cr}^{glob} = 4P_{cr}^{Euler} + 22P_{cr}^{Euler} = 26P_{cr}^{Euler}$

Eq. (6-7): $P_{1,cr}^{glob} = 36P_{cr}^{Euler}$

Eq. (6-8): $P_{1,cr}^{glob} = 15.1 P_{cr}^{Euler} \langle 18 P_{cr}^{Euler}.$

In the above analyses we neglected the elongation (compression) of the columns which takes place during buckling of the frame. This effect can be considered by the Föppl-Papkovich theorem, because we have stiffened the columns against elongation in the foregoing. In the next step we only take the elongation of the columns into account. For the whole frame we thus have

$$I_{elong} = 2A_1 (l/2)^2 \quad \text{(see Fig. 6-1a),}$$

and (assuming again a frame clamped at the bottom and at the top):

$$2P_{1,cr}^{elong} = 4\frac{\pi^2 EI_{elong}}{H^2}. \qquad (6\text{-}10)$$

Hence the "combined" critical load of the frame can be computed from the equation

$$\frac{1}{2P_{1,cr}^{comb}} = \frac{1}{2P_{1,cr}} + \frac{1}{2P_{1,cr}^{elong}}. \qquad (6\text{-}11)$$

It should be mentioned that also multi-bay frames can be analyzed by the method shown, provided they are "proportionate", i.e. the rigidities of the columns and beams correspond to Fig. 6-2, see [Csonka, 1956].

The buckling of frames can also be analyzed by using the analogy between them and sandwich bars. So first we shortly present the main results on buckling of sandwich bars and, second, we establish the analogy.

A sandwich bar with thick faces and antiplane core (Fig. 6-3) is characterized by the following [Allen, 1969]:
   - the faces are Kirchhoff plates;
   - the core
        - is incompressible in the transverse direction,
        - is infinitely soft in the longitudinal direction,
        - has a finite shear rigidity.
The global (overall) bending stiffness of the bar is

$$B_0 = \frac{d^2 t b}{2} E, \qquad (6\text{-}12)$$

the local bending stiffness is

$$B_l = 2\frac{t^3 b}{12}E,$$ (6-13)

and the shear stiffness is

$$S \approx Gbc.$$ (6-14)

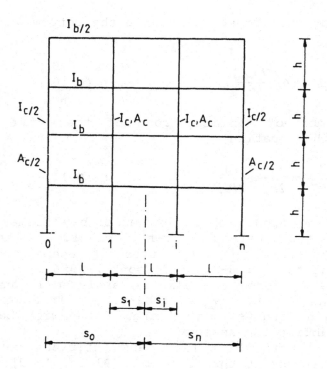

Fig. 6-2 Proportionate frame

With these the following "partial" critical loads can be defined:

$$P_0 = \frac{\pi^2 B_0}{l_0^2},$$ (6-15)

$$P_l = \frac{\pi^2 B_l}{l_0^2},$$ (6-16)

$$P_s = S,$$ (6-17)

see also Eq. (4-3).

From these "partial" loads the actual critical load P_{cr} of the bar can be constructed as follows:

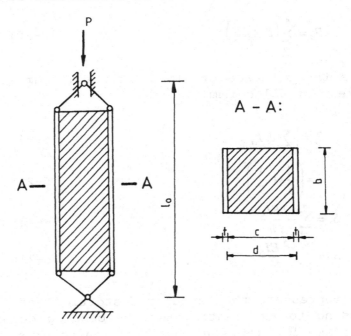

Fig. 6-3 Sandwich bar

Reasoning in the terms of the energy method, we can state that P_0 performs work only with the bending component of deformation, and P_s only with the shearing part of deformation, while P_l performs work with the total deformation. Hence P_0 and P_s combine according to the Föppl-Papkovich theorem, yielding

$$P_{0,s} = \left(\frac{1}{P_0} + \frac{1}{P_s} \right)^{-1},$$ (6-18)

while P_l combines with $P_{0,s}$ according to Southwell's theorem:

$$P_{cr} = P_{0,s} + P_l.$$ (6-19)

Now we have to establish the analogy between frames and sandwich bars [Hegedűs and Kollár, 1987].

Let us confine our treatment to proportionate frames only, according to Fig. 6-2. In this case

$$B_0 = \sum_{i=1}^{n} \left(E A_{ci} \cdot s_i^2 \right)$$ (6-20)

with s_i as the distance of the ith column from the centroid of the areas of all columns,

$$B_l = \left(\sum_{i=1}^{n} E I_{ci} \right),$$ (6-21)

$$S = \left(\frac{1}{S_b} + \frac{1}{S_c} \right)^{-1}$$ (6-22)

with $$S_b = \sum_{i=1}^{n} \frac{12 E I_{bi}}{l \, h}$$ (6-23)

and $$S_c = \sum_{i=1}^{n} \frac{\pi^2 E I_{ci}}{h^2}.$$ (6-24)

Here S_b represents the "angular distortion" of the columns corresponding to the (antisymmetric) bending deformation of the beams, and S_c shows the "sway" of the columns between the (discrete) beams. These two shearing-type deformations have to be added, consequently the inverse values of the corresponding stiffnesses have also to be added, according to the Föppl-Papkovich theorem.

It was shown that in the case of frames with four or more stories the error of the approximate formula is less than 5 %.

7. Buckling of tied arches

The buckling of tied arches with hangers (Fig. 7-1) can be conveniently analyzed in the following way:

In the first step we "smear out" the hangers, i.e. instead of discrete bars we use a "suspension veil". The (overall) buckling of the arch can then be analyzed in a comparatively simple way, see e.g. [Pflüger, 1964], [Kovács, 1975]. This critical load of the arch is denoted by $H_{cr}^{overall}$.

In the second step we consider the buckling of the arch between the discrete hangers, i.e. of a continuous bar on fixed supports, and we obtain H_{cr}^{local}.

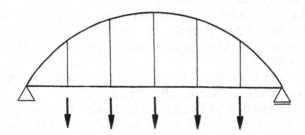

Fig. 7-1   Tied arch

According to the conjecture described in Sect. 3.3 of the chapter " The Southwell and Dunkerley Theorems", the actual ("combined") critical load, H_{cr}, can be obtained from the equation

$$\frac{1}{H_{cr}} = \frac{1}{H_{cr}^{overall}} + \frac{1}{H_{cr}^{local}}.$$ (7-1)

Investigations show that Eq. (7-1) yields a close approximation for tied arches, but for arches with a stiffening beam it gives rather conservative results.

8. Stiffening of buildings against buckling

The load-bearing structure of a building very often does not have enough lateral stiffness to prevent overall buckling of the whole building. To provide stability in these cases, a separate structure is needed, which we will

call the "stiffening core". This core can be a closed
elevator and staircase shaft, but can also consist of
several stiffening walls. In the latter case these walls can
be substituted for by a "fictitious core", whose shear
centre and stiffnesses are identical to those of the wall
system.

In the following we will assume that we have a (real or
fictitious) stiffening core, and we are looking for the
critical load of the building stiffened by this core.

We make the following assumptions:

a) The load-bearing skeleton has zero lateral stiffness
(i.e. the nodes of the frame are pin-joints);

b) The vertical load acting on the core is very small in
comparison to the whole load of the building;

c) The floors of the building are infinitely rigid in
their own planes;

d) The load acting on the building can be considered as
uniformly distributed along the height.

The core-stiffened building may lose its stability in two
directions by bending, and by torsion. These buckling modes
may also combine.

Assumptions a) and b) mean that the load-bearing and the
stiffening functions are provided by separate structural
components. Assumption c) ensures identical deformation of
the two parts. Consequently, we can use the well-known
results of stability of axially loaded bars, but, on the one
hand, the quantities related to stiffnesses have to be taken
from the data of the core, while the quantities related to
loading from the data of the load-bearing structure. On the
other hand, the critical load has to be computed assuming a
perfectly elastic core, since the load is taken not by the
core but by the skeleton, and thus no plasticization of the
core takes place.

On the basis of these considerations we can state the
following:

The critical load, associated with buckling by bending, of
a core-stiffened skeleton, is equal to the (elastic)
critical load of the core. Thus, the critical load of the
simple structure of Fig. 8-1 is equal to

$$N_{cr} = \frac{\pi^2 EI_{core}}{4H^2},$$

(8-1)

and in the case of Fig. 8-2

$$N_{cr} = \left(\sum N_1 \right)_{cr} \approx \left(pH \right)_{cr} = \frac{7.84\,EI_{core}}{h^2},$$ (8-2)

in both directions.

In the case of torsional buckling the phenomenon becomes more complicated.

The critical load, associated with (pure) torsional buckling, of a centrally compressed bar having only torsional rigidity, GI_t, is [Timoshenko and Gere, 1961]:

$$N_{cr} = \frac{GI_t}{i_p^2},$$ (8-3)

where i_p is the radius of gyration (polar radius of inertia), referred to the shear centre of the cross section of the bar.

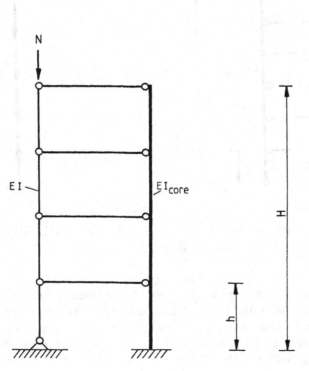

Fig. 8-1   Simple model of a stiffened building loaded on top

The derivation of Eq. (8-3) shows that i_p is related to the loads and to the loaded area as well. Consequently, in our case i_p should be the polar radius of inertia of the loaded area, i.e. of the ground plan of the whole building (provided the loads can be considered as uniformly distributed over the ground plan of the building). In addition, i_p has to be referred to the shear centre of the core.

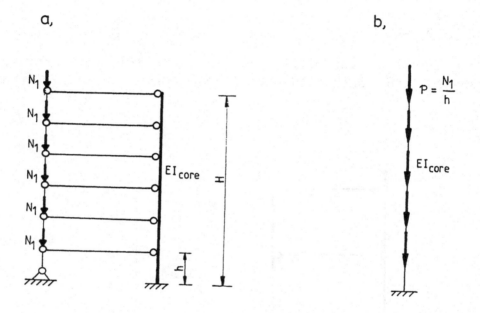

Fig. 8-2   Simple model of a stiffened building loaded
by distributed forces

It follows now that the critical load of torsional buckling depends not only on the torsional stiffness of the core, but also on the magnitude of the loaded area, and on the position of the core in relation to this area. A core close to the centroid of the ground plan of the building is thus much more effective than a core in one corner of the ground plan.

Formula (8-3) is valid for a column loaded on top. If the load of the column is uniformly distributed along the height, Eq. (8-3) gives the critical value of the maximum compressive force acting at the lower end of the column.

As can be seen, these results are analogous to the buckling by shear deformation, discussed in Sect. 4. The buckling modes are also similar if we take into account the fact that in the present case the angle of twist per unit length corresponds to the angular distortion of a bar buckling by shear. Hence the buckling mode of a column under constant compressive force is also here arbitrary, while under distributed load the buckling mode corresponds to Fig. 4-2b (with the angle of twist instead of displacement y on the abscissa).

Let us now investigate the pure torsional buckling of a building stiffened by a core which has only warping rigidity (EI_ω). We can again use the results of a compressed bar [Timoshenko and Gere, 1961], whose differential equation has the same form as for buckling by bending. Hence we obtain for a column loaded on top:

$$N_{cr} = \frac{1}{i_p^2} \frac{\pi^2 EI_\omega}{4 H^2}$$ (8-4)

and for a column under distributed load:

$$N_{cr} = \frac{1}{i_p^2} \frac{7.84 EI_\omega}{H^2}$$ (8-5)

In the latter case N_{cr} means the maximum value of the compressive force, arising at the bottom. In these formulas again i_p stands for the polar radius of inertia of the loaded area.

With knowledge of these results we can easily construct the critical load of a building under distributed load in pure torsional buckling, if the core has both torsional and warping rigidities. We have to use Southwell's theorem:

$$N_{cr}^{tors} = \frac{1}{i_p^2}\left(GI_t + \frac{7.84\,EI_\omega}{H^2} \right).$$ (8-6)

Our last problem is the combination of bending and torsional buckling modes.

The theory of torsional buckling of compressed bars [Timoshenko and Gere, 1961] shows that this combination comes about if the centroid of the cross section (i.e. of the loaded area) does not coincide with the shear centre. If the centroid lies on one of the principal axes of inertia, drawn across the shear centre, then buckling by bending perpendicularly to this axis combines with torsional buckling; and if the centroid does not lie on any of the principal axes, then buckling by bending in both directions combine with torsional buckling.

The three possible situations are shown is Fig. 8-3. In the case of Fig. 8-3a there is no combination. In the cases of Fig. 8-3b and c, the critical load of buckling by bending perpendicularly to the principal axis x $\left(N_{cr}^{x}\right)$ combines with N_{cr}^{tors}, and, using the Föppl- Papkovich theorem, we obtain

$$ N_{cr} = \left(\frac{1}{N_{cr}^{x}} + \frac{1}{N_{cr}^{tors}} \right)^{1}. \qquad (8-7) $$

while for Fig. 8-3d we have

$$ N_{cr} = \left(\frac{1}{N_{cr}^{x}} + \frac{1}{N_{cr}^{y}} + \frac{1}{N_{cr}^{tors}} \right)^{1} \qquad (8-8) $$

The formulas (8-7) and (8-8) may deviate considerably from the exact result and furnish a critical load on the safe side. The exact results can be found in [Zalka and Armer, 1992].

9. Lateral stability of beams

The summation theorems are particularly useful for computing the critical loads of beams which lose their stability by lateral buckling. This problem is rather complicated in the sense that there are many factors influencing this phenomenon, and their combinations yield a very high number of different cases which should be solved separately. On the other hand, if we make use of the summation theorems, every case can be constructed from a few simple, basic cases.

The factors influencing lateral buckling are the following:

- the support conditions of the beam ("fork-like" supports, i.e. simple supports which prevent rotation; clamping; suspension; with or without cantilevers protruding beyond the supports etc.),
- the cross section (thin-walled or solid, symmetric or not, constant or variable along the length etc.),
- the load (uniformly distributed or concentrated etc.),
- the position of the load in relation to the shear centre (above or underneath etc.).
In the following we will assume that the cross section of the beam is thin-walled, monosymmetric, and constant along the length, furthermore that the load acts in the plane of symmetry of the cross section.

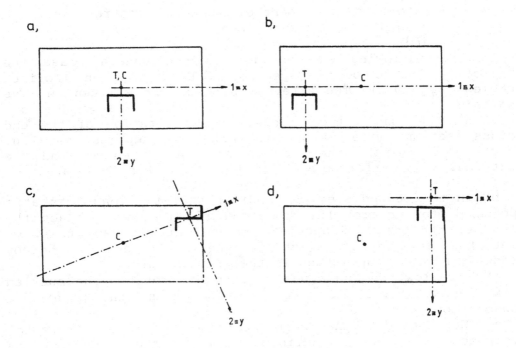

Fig. 8-3 Various cases of combination of bending
and torsional buckling modes

We exclude from our treatment the so-called shell-beams with cross sections according to Fig. 9-1, because their lateral buckling analysis needs special considerations, see in [Kollár, 1973].

For the treatment of the subject it is most expedient to apply the energy method.

The internal work (strain energy) of a laterally buckled bar can be written as

$$L_i = \frac{GI_t}{2} \int_0^l (\phi')^2 \, dz + \frac{EI_\omega}{2} \int_0^l (\phi'')^2 \, dz + \frac{EI_y}{2} \int_0^l (u_T'')^2 \, dz \qquad (9\text{-}1)$$

[Chwalla, 1944], [Bleich, 1952], with the notations

GI_t - torsional rigidity,
EI_ω - warping rigidity,
EI_y - lateral bending rigidity,
ϕ - angle of twist,
u_T - lateral displacement of the shear centre,
z - co-ordinate along the beam axis,
'   - d/dz.

In the following we consider as the general case the uniformly loaded beam suspended at both ends on inclined cables (Fig. 9-2). The external work can be written in five parts:

a) In the first step we prevent the rotation of the end cross sections (i.e. we assume fork-like supports, see Fig. 9-3), we transpose the load to the shear centre, and the external work performed will be determined for this case.

b) In the second step we consider that the load acts at a point different from the shear centre, i.e. we let act the load at its point of application and, at the same time, we let the reverse load act at the shear centre, hereby cancelling the load acting in the previous step.

c) We allow the end cross sections to rotate; the buckled beam hence rotates as a rigid body around the points of suspension.

d) We consider the work performed by the horizontal components of the inclined cable forces during the deformation described in step a).

e) Since the points of suspension do not coincide with the centroid, the compressive forces described under d) act eccentrically on the beam. This bending moment also performs external work with the deformation under a).

All these five parts of external work have the form

$$L_{e,k} = q\,L_{e,k}^{(1)} \qquad (k = 1,2...5) \qquad\qquad (9\text{-}2)$$

with $L_{e,k}^{(1)}$ as the external work performed by the load q=1.

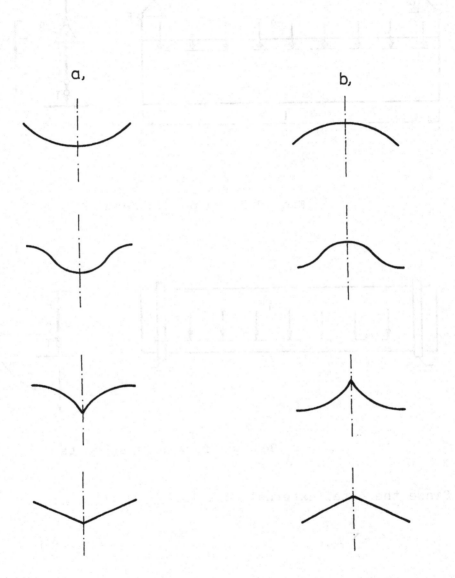

a, b,

Fig. 9-1 Cross sections of shell beams

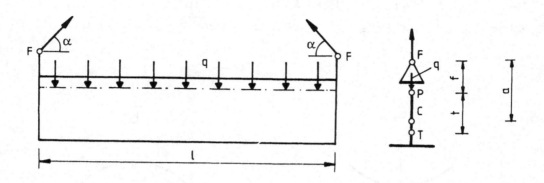

Fig. 9-2   Suspended beam

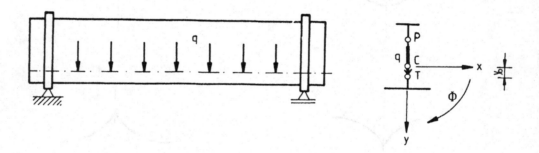

Fig. 9-3   Beam on fork-like supports

Since the total external work is

$$L_e = \sum_{k=1}^{5} L_{e,k},$$
(9-3)

so, equating L_i (9-1) with L_e:

$$L_i = q \sum_{k=1}^{5} L_{e,k}^{(1)},$$
(9-4)

or

$$\frac{1}{q} = \frac{\sum\limits_{k=1}^{5} L_{e,k}^{(1)}}{L_i} = \sum\limits_{k=1}^{5} \frac{1}{q_k} \tag{9-5}$$

with

$$q_k = \frac{L_i}{L_{e,k}}, \tag{9-6}$$

as the (partial) critical load of the kth case.

Eq. (9-5) shows that the inverse of the critical load of the beam can be obtained as the sum of the inverses of the partial critical loads. Since in the five cases we prevented some deformations of the beam, Eq. (9-5) can be considered a Föppl-Papkovich theorem.

We compute each one of the partial critical loads by assuming a buckling mode pertaining to the case in question. Since these do not coincide with the true buckling mode of the suspended beam, we obtain partial critical loads inferior to those to be obtained with the true, unique buckling mode of the beam (which we do not know), and thus we arrive at a lower critical load than the exact value, as is conformable to the theory of summation theorems.

However, since we determine the partial critical loads by the energy method which gives, as a rule, higher results than the exact ones, we can expect that these two opposite errors (at least to a certain degree) extinguish each other.

For the five particular cases mentioned before we can write the expressions for the external work as follows:

a) For the simply supported beam on fork-like supports we can use the result of Meissner [1955]:

$$0.01218 \frac{l^4}{EI_y} \left(q_{cr}^{fork-like}\right)^2 - 0.5725 \beta_1 \, q_{cr}^{fork-like} - \frac{\pi^2}{l^2}\left(GI_t + \frac{\pi^2}{l^2} EI_\omega\right) = 0. \tag{9-7}$$

where

$$\beta_1 = J_1 + J_2 - 2x_0, \tag{9-8}$$

with

$$J_1 = \frac{\int\limits_{(A)} y^3 \, dA}{I_x} \tag{9-9a}$$

and

$$J_2 = \frac{\int\limits_{(A)} x^2 y \, dA}{I_x},$$ (9-9b)

with I_x as the moment of inertia referred to the horizontal axis x, see Fig. 9-3.

Meissner's formula, which took only one term of each deformation function into account, gives a very good approximation provided that EI_y is not too great or $\dfrac{EI_\omega \pi^2}{l^2}$ is not too small in comparison with GI_t.

b) For the beam loaded at the centroid and at the shear centre by opposite forces we can easily find the critical load intensity:

$$q_{cr}^{opp} = \frac{\pi^2}{t\,l^2}\left(GI_t + \frac{\pi^2}{l^2} EI_\omega\right),$$ (9-10)

see Fig. 9-2.

c) For the beam rotating about the points of suspension we find

$$q_{cr}^{susp} = \frac{12.176\, f}{\dfrac{l^4}{\pi^2 EI_y} + \dfrac{t^2 l^2}{GI_t + \dfrac{\pi^2}{l^2} EI_\omega}}$$ (9-11)

see Fig. 9-2.

d) Under the compressive force N the beam buckles sideways by combined torsional buckling and buckling by bending. The critical value of N is given by the smaller root of the quadratic equation (see [Timoshenko and Gere, 1961]):

$$\frac{i_p^2}{i_p^2 + y_0^2} N_{cr}^2 - \left(N_{y,cr} + N_{\phi,cr}\right) N_{cr} + N_{y,cr} N_{\phi,cr} = 0.$$ (9-12)

with

$$i_p^2 = \frac{I_x + I_y}{A},$$ (9-13)

$$N_{y,cr} = \frac{\pi^2 EI_y}{l^2},$$ (9-14a)

and

$$N_{\phi,cr} = \frac{GI_t + \frac{\pi^2}{l^2} EI}{i_p^2 + y_0^2},$$ (9-14b)

see Fig. 9-3.

From N_{cr} we obtain the critical load intensity:

$$q_{cr}^N = \frac{2N_{cr}}{l \cot \alpha}.$$ (9-15)

e) Finally, the critical value of the bending moment $M_N = Na$ caused by the eccentricity of the compressive force N (Fig. 9-2) is given by the positive root of the equation [Timoshenko and Gere, 1961]:

$$\frac{1}{EI_y} M_{cr}^2 - \frac{\pi^2}{l^2} \beta_1 M_{cr} - \frac{\pi^2}{l^2} \left(GI_t + \frac{\pi^2}{l^2} EI_\omega \right) = 0.$$ (9-16)

and the critical load is

$$q_{cr}^M = \frac{2M_{cr}}{al \cot \alpha}.$$ (9-17)

When summing up the inverse values of the five critical loads we have to observe that if one of these partial critical loads have a negative sign (e.g. in case b), if the load acts beneath the shear centre), then we have to consider this load as infinitely great (i.e. its inverse value should be taken equal to zero), because otherwise we cannot be sure whether the result of the summation theorem lies on the safe side or not.

REFERENCES

ALLEN,H.G. 1969. Analysis and Design of Structural Sandwich Panels. Pergamon Press, Oxford etc.

BLEICH,FR. 1952. Buckling Strength of Metal Structures. McGraw-Hill Book Comp., New York.

CHWALLA,E. 1944. Kippung von Trägern mit einfachsymmetrischen, dünnwandigen und offenen Querschitten. Sitzungsbereichte der Akademie der Wissenschaften in Wien. Abt. IIa, Bd. 153, 25-60.

CSONKA,P. 1956. Über proportionierte Rahmen. Die Bautechnik 33, 19-20.

CSONKA,P. 1961. Buckling of bars elastically built-in along their entire length. Acta Techn. Hung. 32, 424-427.

HEGEDÜS,I. and KOLLÁR,L.P.
        1987. Stabilitätsuntersuchung von Rahmen und Wandscheiben mit der Sandwichtheorie. Bautechnik, 64, 420-425.

KÁRMÁN, Th. v. and BIOT,M.A.
        1940. Mathematical methods in engineering. McGraw-Hill, New York.

KOLLÁR,L. 1973. Statik und Stabilität der Schalenbogen und Schalenbalken. Akadémiai Kiadó, Budapest - W.Ernst und Sohn. Berlin etc.

KOVÁCS,I. 1975. Allgemeines Knicken des mit Hängestangen und Balken versteiften Parabelbogens. Doctor's thesis. Technische Universität Stuttgart.

MEISSNER,F.1955. Einige Auswertungsergebnisse der Kipptheorie einfachsymmetrischer Balkenträger. Der Stahlbau, 24, 110-113.

PFLÜGER,A. 1964. Stabilitätsprobleme der Elastostatik. 2. Aufl. Springer, Berlin etc.

TIMOSHENKO,S.P. and GERE,J.M.
        1961. Theory of Elastic Stability. McGraw-Hill, New York etc.

ZALKA,K.A. and ARMER,G.S.T.
        1992. Stability of large structures. Butterworth-Heinemann, Oxford.

Printed in the United States
By Bookmasters